National Security Implications of Climate Change for U.S. Naval Forces

Committee on National Security Implications of Climate Change
for U.S. Naval Forces

Naval Studies Board

Division on Engineering and Physical Sciences

NATIONAL RESEARCH COUNCIL
OF THE NATIONAL ACADEMIES

THE NATIONAL ACADEMIES PRESS
Washington, D.C.
www.nap.edu

THE NATIONAL ACADEMIES PRESS 500 Fifth Street, N.W. Washington, DC 20001

This study was supported by Contract No. N00014-05-G-0288, DO #25 between the National Academy of Sciences and the Department of the Navy. Any opinions, findings, conclusions, or recommendations expressed in this publication are those of the author(s) and do not necessarily reflect the views of the organizations or agencies that provided support for the project.

International Standard Book Number-13: 978-0-309-15425-3
International Standard Book Number-10: 0-309-15425-1

Copies of this report are available from:

Naval Studies Board, National Research Council, The Keck Center of the National Academies, 500 Fifth Street, N.W., Room WS904, Washington, DC 20001; and

The National Academies Press, 500 Fifth Street, N.W., Lockbox 285, Washington, DC 20055; (800) 624-6242 or (202) 334-3313 (in the Washington metropolitan area); Internet, http://www.nap.edu.

Printed in the United States of America

THE NATIONAL ACADEMIES

Advisers to the Nation on Science, Engineering, and Medicine

The **National Academy of Sciences** is a private, nonprofit, self-perpetuating society of distinguished scholars engaged in scientific and engineering research, dedicated to the furtherance of science and technology and to their use for the general welfare. Upon the authority of the charter granted to it by the Congress in 1863, the Academy has a mandate that requires it to advise the federal government on scientific and technical matters. Dr. Ralph J. Cicerone is president of the National Academy of Sciences.

The **National Academy of Engineering** was established in 1964, under the charter of the National Academy of Sciences, as a parallel organization of outstanding engineers. It is autonomous in its administration and in the selection of its members, sharing with the National Academy of Sciences the responsibility for advising the federal government. The National Academy of Engineering also sponsors engineering programs aimed at meeting national needs, encourages education and research, and recognizes the superior achievements of engineers. Dr. Charles M. Vest is president of the National Academy of Engineering.

The **Institute of Medicine** was established in 1970 by the National Academy of Sciences to secure the services of eminent members of appropriate professions in the examination of policy matters pertaining to the health of the public. The Institute acts under the responsibility given to the National Academy of Sciences by its congressional charter to be an adviser to the federal government and, upon its own initiative, to identify issues of medical care, research, and education. Dr. Harvey V. Fineberg is president of the Institute of Medicine.

The **National Research Council** was organized by the National Academy of Sciences in 1916 to associate the broad community of science and technology with the Academy's purposes of furthering knowledge and advising the federal government. Functioning in accordance with general policies determined by the Academy, the Council has become the principal operating agency of both the National Academy of Sciences and the National Academy of Engineering in providing services to the government, the public, and the scientific and engineering communities. The Council is administered jointly by both Academies and the Institute of Medicine. Dr. Ralph J. Cicerone and Dr. Charles M. Vest are chair and vice chair, respectively, of the National Research Council.

www.national-academies.org

COMMITTEE ON NATIONAL SECURITY IMPLICATIONS OF CLIMATE CHANGE FOR U.S. NAVAL FORCES

FRANK L. BOWMAN, Strategic Decisions, LLC, *Co-Chair*
ANTONIO J. BUSALACCHI, JR., University of Maryland, *Co-Chair*
ARTHUR B. BAGGEROER, Massachusetts Institute of Technology
CECILIA M. BITZ, University of Washington
SHARON E. BURKE, Center for a New American Security (through June 22, 2010)
RONALD FILADELFO, Center for Naval Analyses
JEFFREY M. GARRETT, Mercer Island, Washington
HARRY W. JENKINS, JR., Gainesville, Virginia
CATHERINE M. KELLEHER, University of Maryland and Brown University
MAHLON C. KENNICUTT II, Texas A&M University
RONALD R. LUMAN, Applied Physics Laboratory, Johns Hopkins University
W. BERRY LYONS, Ohio State University
JAMES J. McCARTHY, Harvard University
MICHAEL J. McPHADEN, National Oceanic and Atmospheric Administration
JOHN H. MOXLEY III, Solvang, California
DAVID J. NASH, Dave Nash & Associates, LLC
HEIDI C. PERRY, Charles Stark Draper Laboratory, Inc.
J. MARSHALL SHEPHERD, University of Georgia
CHARLES F. WALD, Deloitte Services, LLP
DAVID A. WHELAN, The Boeing Company
CARL WUNSCH, Massachusetts Institute of Technology

Staff

CHARLES F. DRAPER, Director, Naval Studies Board
BILLY M. WILLIAMS, Study Director
RAYMOND S. WIDMAYER, Senior Program Officer
MARTA V. HERNANDEZ, Associate Program Officer
SUSAN G. CAMPBELL, Administrative Coordinator
MARY G. GORDON, Information Officer
SEKOU O. JACKSON, Senior Program Assistant
SIDNEY G. REED, JR., Consultant

NAVAL STUDIES BOARD

Staff

CHARLES F. DRAPER, Director
RAYMOND S. WIDMAYER, Senior Program Officer
BILLY M. WILLIAMS, Senior Program Officer
MARTA V. HERNANDEZ, Associate Program Officer
SUSAN G. CAMPBELL, Administrative Coordinator
MARY G. GORDON, Information Officer
SEKOU O. JACKSON, Senior Program Assistant
SIDNEY G. REED, JR., Consultant

Preface

A 2007 report by the CNA Corporation, based on a study conducted by former flag and general officers, concluded that worldwide climate effects could have major consequences for the military.[1] In testimony provided before the Senate Committee on Foreign Relations in May 2007, it was noted that "projected climate change poses a serious threat to America's national security; climate change acts as a threat multiplier for instability in some of the most volatile regions of the world; projected climate change will add to tensions even in stable regions of the world; and climate change, national security and energy dependence are a related set of global challenges."[2] More recently the 2010 *Quadrennial Defense Review* (QDR), conducted internally by the Department of Defense (DOD) to identify military capabilities that could contribute to fulfilling U.S. national security needs, stated that "climate change and energy will play significant roles in the future security environment."[3]

The National Academies[4] has undertaken a number of completed and recently initiated activities aimed at examining climate change. In particular, in response to Public Law 110-161, the National Academies is undertaking a series of coor-

[1]Military Advisory Board. 2007. *National Security and the Threat of Climate Change*, CNA Corporation, Alexandria, Va.

[2]Testimony of ADM Joseph Prueher, USN (Ret.), Member, Military Advisory Board, Center for Naval Analyses Corporation report *National Security and the Threat of Climate Change,* before the Committee on Foreign Relations, U.S. Senate, May 9, 2007.

[3]Secretary of Defense (Robert M. Gates). 2010. *Quadrennial Defense Review,* Department of Defense, Washington, D.C., February, p. xv.

[4]The National Academies comprises the National Academy of Sciences, the National Academy of Engineering, the Institute of Medicine, and the National Research Council.

dinated activities—known collectively as America's Climate Choices (ACC)—to study the serious and sweeping issues associated with global climate change, including the science and technology challenges involved. The ACC studies are an effort to provide advice on the most effective steps and most promising strategies that can be taken to respond to global climate change. In short, the ACC suite of activities will produce a broad, action-oriented, and authoritative set of analyses to inform and guide responses to climate change across the nation.[5]

Outside of the National Academies, a number of organizations are likewise examining climate change, most notably the Intergovernmental Panel on Climate Change, or IPCC, which was formed in 1988 under the auspices of the United Nations. Since its inception, the IPCC's function has been to provide assessments of the science of climate change. In 2007, it released its *Fourth Assessment Report* (AR4), which noted among its many findings that "warming of the climate system is unequivocal, as is now evident from observations of increase in global average air and ocean temperatures, widespread melting of snow and ice, and rising global mean sea level."[6]

Accordingly, what does climate change mean for the U.S. naval forces (i.e., the U.S. Navy, Marine Corps, and Coast Guard); and, specifically, what does climate change mean for U.S. naval forces in terms of the national security implications? To understand this question in greater depth and in conjunction with the Navy's efforts with respect to the 2010 Quadrennial Defense Review, the Office of the Chief of Naval Operations (CNO) tasked the Center for a New American Security (CNAS) and CNA to undertake separate studies.[7] The CNAS study, recently completed after a 3-month duration, was aimed at examining the national security implications of climate change through a strategic lens. The CNA study, completed in 2009 after a 6-month duration, focused on the operational implications for the Navy as a result of increased maritime activity in the Arctic region.[8]

---

[5]The initial four reports from the America's Climate Choices (ACC) studies are these: National Research Council, 2010, *Advancing the Science of Climate Change, Informing an Effective Response to Climate Change, Limiting the Magnitude of Future Climate Change,* and *Adapting to the Impacts of Climate Change,* The National Academies Press, Washington, D.C. Additional information on the ACC studies is available at http://americasclimatechoices.org. Accessed June 4, 2010.

[6]See Intergovernmental Panel on Climate Change, 2007, "Climate Change 2007: The Physical Science Basis," Working Group I contribution to the *Fourth Assessment Report of the Intergovernmental Panel on Climate Change* (Susan Solomon, Dahe Qin, Martin Manning, Zhenlin Chen, Melinda Marquis, Kristen B. Averyt, Melinda M.B. Tignor, and Henry LeRoy Miller [eds.]), Cambridge University Press, Cambridge, United Kingdom, and New York.

[7]During the CNO Executive Board meeting in May 2009, the CNO directed the Oceanographer of the Navy to establish and lead Task Force Climate Change in order to develop a comprehensive approach to guide the Navy's future public, strategic, and policy discussions. See Vice Chief of Naval Operations (ADM Jonathan W. Greenert, USN), Memorandum 4000 Ser N09/9U103035, "Task Force Climate Change Charter," Washington, D.C., October 30, 2009.

[8]See Center for a New American Security, 2008, *Uncharted Waters, The U.S. Navy and Navigating Climate Change,* Washington, D.C., December; and Michael D. Bowes, 2009, *Impact of Climate*

In a letter to the president of the National Academy of Sciences, the CNO requested that the Naval Studies Board (NSB) begin a new critical-area study in fiscal year (FY) 2009 examining the global implications of climate change for the naval services.[9] In essence, the NSB study would follow on the heels of the CNAS and CNA reports. It would leverage these and other study insights along with the expertise and experience of the National Academies in the area of climate change. Moreover, certain areas from the CNAS and CNA reports either have not been considered or have not been explored in depth because of study duration. These areas are included in the investigation by the NSB in order to provide a thorough assessment of the national security implications of climate change for U.S. naval forces.

TERMS OF REFERENCE

The letter dated September 12, 2008, from ADM Gary Roughead, Chief of Naval Operations, to Dr. Ralph J. Cicerone, President of the National Academy of Sciences, requested that the National Research Council's (NRC's) Naval Studies Board (NSB) conduct a comprehensive study on the national security implications of climate change for U.S. naval forces (i.e., the U.S. Navy, Marine Corps, and Coast Guard), based on the Intergovernmental Panel on Climate Change assessments and other subsequent relevant literature.[10]

Accordingly, the NRC, under the auspices of its NSB, established the Committee on National Security Implications of Climate Change for U.S. Naval Forces in September 2009.[11] The study's terms of reference, formulated by the CNO's staff in consultation with the NSB chair and director, charge the committee to produce two reports over a 15-month period. The first report produced by the committee was a letter report that summarized the immediate challenges in each of the four areas of the terms of reference (see Box P.1). Specifically, the letter report highlighted issues brought to the committee's attention during its first three meetings that could have potential near-term impacts, impose a need for near-term awareness, or require near-term planning to ensure that longer-term naval capabilities are protected. The requested letter report was delivered to the CNO and other naval leadership stakeholders in April 2010.[12]

After completing its letter report and conducting additional data gathering, the committee was requested to produce a comprehensive final report that addresses the full terms of reference. This report—the committee's second and

---

Change on Naval Operations in the Arctic, CAB D0020034.A3/1REV, Center for Naval Analyses, Alexandria, Va., April.

[9]Letter of request from ADM G. Roughead, USN, Chief of Naval Operations, to Dr. Ralph J. Cicerone, President, National Academy of Sciences, September 12, 2008.

[10]Ibid.

[11]Biographical information for the committee and staff is presented in Appendix C.

[12]The committee's letter report is provided in Appendix D.

BOX P.1
Terms of Reference—National Security Implications of Climate Change for U.S. Naval Forces

At the request of the Chief of Naval Operations, the Naval Studies Board of the National Research Council will establish a committee to study the national security implications of climate change for U.S. naval forces (i.e., the U.S. Navy, Marine Corps, and Coast Guard). Based on the Intergovernmental Panel on Climate Change assessments and other subsequent relevant literature reviewed by the committee, the study will:

1. Examine the potential impact on U.S. future naval operations and capabilities as a result of climate change (e.g., how will U.S. future naval operations be impacted and what capabilities will be needed for U.S. future naval forces as a result of climate change? This includes an assessment of the U.S. Coast Guard and Marine Corps, and where the U.S. Navy might be required to supplement or augment their capabilities).
2. Assess the robustness of the Department of Defense's infrastructure for supporting U.S. future naval operations and capabilities in the context of potential climate change impacts (e.g., are there any U.S. military installations and/or forward-deployed bases providing support to U.S. naval forces that are potentially vulnerable as a result of climate change?).
3. Determine the potential impact climate change will have on allied force operations and capabilities (e.g., are there any allies who may need U.S. naval force support as a result of climate change? Conversely, which allied force operations and capabilities may U.S. naval forces wish to leverage as a result of climate change?).
4. Examine the potential impact on U.S. future naval antisubmarine warfare operations and capabilities in the world's oceans as a result of climate change; specifically, the technical underpinnings for projecting U.S. undersea dominance in light of the changing physical properties of the oceans.

This 15-month study will produce two reports: (1) a letter report following the third full committee meeting that summarizes the immediate challenges for U.S. naval forces in addressing each of the four above areas, as well as recommends approaches to address these challenges; (2) a comprehensive report that addresses in greater depth the full terms of reference.

final report—builds on the near-term challenges identified in the letter report. In total, the committee believes that both reports have responded productively to the CNO's charge.

THE COMMITTEE'S APPROACH

In addressing its charge, the committee studied a range of issues associated with national security implications of climate change for U.S. naval forces. In addition, the committee reviewed IPCC assessments and other subsequent literature, as well as reference documents from scientific and operational communities. The findings and recommendations in this final report are based on wide-ranging input from experts, both internal and external to the Department of the Navy and the Department of Defense, and on the committee's own analysis, which draws on the expertise and experience of its members.

The committee first convened in September 2009. After its third full meeting, the committee drafted its letter report. The committee went on to convene additional meetings and data-gathering sessions over a period of 7 months, both to gather input from the relevant communities and to discuss its findings and recommendations. A summary of the committee's meetings is provided below:

- *September 17-18, 2009, in Washington, D.C.* First full committee meeting: Briefings on current climate change and energy-related initiatives from Navy Task Force Climate Change; Navy Task Force Energy; the Navy Quadrennial Defense Review (QDR) Integration Group; the Office of the Deputy Chief of Naval Operations for Integration of Capabilities and Resources (N81); the Office of Facilities Branch Head, U.S. Marine Corps; the Office of Environmental Management Section, Headquarters, U.S. Marine Corps; and the Office of Policy Integration, Headquarters, U.S. Coast Guard. Additionally the committee received briefings on recently completed climate-change-related reports by the Center for a New American Security, the CNA, and the National Research Council.
- *October 19-20, 2009, in Washington, D.C.* Second full committee meeting: Briefings on climate-change-related national security issues, naval installation vulnerabilities, and current research activities by representatives from the National Intelligence Council, Woods Hole Oceanographic Institution, Oak Ridge National Laboratory, Navy Task Force Climate Change, Naval Installations Command, the Office of Naval Research, the U.S. Geological Survey, the National Ice Center, the National Oceanic and Atmospheric Administration, the University of Washington, and the University of Colorado.
- *November 19-20, 2009, in Washington, D.C.* Third full committee meeting. Briefings on human dimensions, allies' perspectives, water resource issues, and maritime operational perspectives of climate change from Columbia University's Center for International Earth Science Information Network; the Pacific Institute for Studies in Development, Environment, and Security; the British Defence Staff

of the United States British Embassy; the Office of the Deputy Chief of Naval Operations for Information Plans and Strategy; and the Commandant, U.S. Coast Guard.

• *January 7-8, 2010, in Washington, D.C.* Fourth full committee meeting. Briefings on U.S. naval and DOD military prespectives on climate change from Navy Task Force Climate Change, USMC Expeditionary Energy Office, the First Naval Construction Division, Navy QDR Integration Group, and DOD's Strategic Environmental Research Group; allies' prespectives from the Office of Defense Research and Development—Embassy of Canada; climate change initiatives and research coordination from the White House Office of Science and Technology Policy (OSTP); and coastal vulnerability mapping from the U.S. Geological Survey.

• *February 4-5, 2010, in Washington, D.C.* Fifth full committee meeting. Briefings on climate and disease, allies' perspectives, climate science, and U.S. Navy and DOD perspectives from the Assistant Surgeon General, Department of Health and Human Services/Centers for Disease Control and Prevention; the National Academies Institute of Medicine; the Director, General Defence Research and Development—Canada; the United States Joint Forces Command; the Joint Global Change Research Institute; the White House OSTP; the Navy Task Force Climate Change; the Navy QDR Integration Group; and Project MEDEA.

• *February 25, 2010, in Stennis, Mississippi.* Site visit/small group data-gathering session. Briefings on Naval Meteorology and Oceanography Command's climate-change-related capabilities, perspectives, and plans from Naval Oceanographic Office.

• *March 5, 2010, in Washington, D.C.* Site visit/small group data-gathering session. Briefing on NATO and allied partners' perspectives on national security and climate change issues, capabilities, and plans by ADM James G. Stavridis, USN, Commander of the United States European Command, and Supreme Allied Commander, Europe.

• *March 22-23, 2010, in Washington, D.C.* Sixth full committee meeting. Briefings on U.S. Department of State, U.S. naval services, and allies' perspectives on climate change; U.S. naval services humanitarian assistance and disaster relief response in Haiti; and updated climate science and climate model projections. Briefings received from the Defense Attaché—Royal Norwegian Embassy; Office of the Commander, U.S. Second Fleet; Office of the Commander, Naval Installations Command; Office of the Commander, Fourth Fleet, U.S Navy Southern Command; the Director, Plans, Policy and Operations, USMC Future Operations Group; Navy Task Force Climate Change; Project MEDEA; and National Oceanic and Atmospheric Administration/Geophysical Fluid Dynamics Laboratory.

• *April 12-16, 2010, in Irvine, California.* Seventh full committee meeting. Committee deliberations and report drafting.

The months between the committee's last meeting and the publication of the report were spent preparing the draft manuscript, gathering additonal information, reviewing and responding to the external review comments, editing the report, and conducting the security review needed to produce an unclassified report.

The committee believes that it has responded productively to the original tasking by providing in this final report a comprehensive analysis of the primary issues associated with the national security implications of climate change for U.S. naval forces. The committee thanks the many briefers who presented information essential to the writing of this report. In particular, the committee is grateful to CAPT Timothy Gallaudet, USN, Deputy Director, Navy Task Force Climate Change, and CDR Esther McClure, USN, Head, Energy and Environmental Issues, Navy QDR Integration Group (who has, since the writing of this report, retired from the U.S. Navy and is now serving as the strategy action officer in the Office of the Secretary of Defense-Policy)—both of whom helped facilitate the committee's effort in gathering information related to the study.

Acknowledgment of Reviewers

This report has been reviewed in draft form by individuals chosen for their diverse perspectives and technical expertise, in accordance with procedures approved by the National Research Council's Report Review Committee. The purpose of this independent review is to provide candid and critical comments that will assist the institution in making its published report as sound as possible and to ensure that the report meets institutional standards for objectivity, evidence, and responsiveness to the study charge. The review comments and draft manuscript remain confidential to protect the integrity of the deliberative process. We wish to thank the following individuals for their review of this report:

David L. Bradley, Applied Research Laboratory, Pennsylvania State University,
Rita R. Colwell, University of Maryland,
Florence Fetterer, University of Colorado,
Paul G. Gaffney II, VADM, USN (Ret.), Monmouth University,
James D. Hull, VADM, USCG (Ret.), Annapolis, Maryland,
William A. LaPlante, Applied Physics Laboratory, Johns Hopkins University,
Pamela A. Matson, Stanford University,
Joseph Pedlosky, Woods Hole Oceanographic Institution, and
John E. Rhodes, LtGen, USMC (Ret.), Balboa, California.

Although the reviewers listed above provided many constructive comments and suggestions, they were not asked to endorse the conclusions or recommendations, nor did they see the final draft of the report before its release. The review of this report was overseen by Robert A. Frosch, Harvard University. Appointed by the National Research Council, he was responsible for making certain that an independent examination of this report was carried out in accordance with institutional procedures and that all review comments were carefully considered. Responsibility for the final content of this report rests entirely with the authoring committee and the institution.

Contents

APPENDIXES

Summary

In response to the Chief of Naval Operations (CNO), the National Research Council appointed a committee operating under the auspices of the Naval Studies Board to study the national security implications of climate change for U.S. naval forces. In conducting its study, the committee found that even the most moderate current trends in climate, if continued, will present new national security challenges for the U.S. Navy, Marine Corps, and Coast Guard.[1] While the timing, degree, and consequence of future climate change impacts remain uncertain, many changes are already under way in regions around the world, such as in the Arctic, and call for action by U.S. naval leadership in response.

The terms of reference (TOR) directed that the study be based on Intergovernmental Panel on Climate Change (IPCC) scenarios and other peer-reviewed assessments. Therefore, the committee did not address the science of climate change or challenge the scenarios on which the committee's findings and recommendations are based. The TOR directed the study to:

1. Examine the potential impact on U.S. future naval operations and capabilities as a result of climate change. . . .
2. Assess the robustness of the Department of Defense's infrastructure for supporting U.S. future naval operations and capabilities in the context of potential climate change impacts. . . .
3. Determine the potential impact climate change will have on allied force operations and capabilities. . . .
4. Examine the potential impact on U.S. future naval antisubmarine warfare

[1]Throughout this report, the terms "Navy," "Marine Corps," and "Coast Guard" are used. Unless stated otherwise, these refer to the "U.S. Navy," "U.S. Marine Corps," and "U.S. Coast Guard."

operations and capabilities in the world's oceans as a result of climate change; specifically, the technical underpinnings for projecting U.S. undersea dominance in light of the changing physical properties of the oceans.

This final report addresses both the near- and long-term implications for U.S. naval forces in each of the four areas of the TOR, and provides corresponding findings and recommendations.[2,3] In an effort to identify areas that need action by U.S. naval leadership, this report and its findings and recommendations are organized around six discussion areas—all presented within the context of a changing climate.

1. Disputes over boundaries and exclusive economic zones as a result of new maritime transits and competition for new resources;
2. Strains on naval capabilities—given continuing first responder missions, and the opening of new international and territorial waters;
3. Vulnerabilities to naval coastal installations due to sea-level rise and increased storm surges;
4. Demands for establishing greater U.S., allied, and/or international maritime partnerships;
5. Impacts on the technical underpinnings that enable, in part, naval force capabilities, particularly those that operate and train in the Arctic; and
6. Investments for additional research and development that have implications for future naval operations and capabilities and might not be met by other groups pursuing climate-related research.

In total, the conclusions from this study can be viewed in the context of six areas for action by U.S. naval leadership. These conclusions, along with their corresponding major findings and recommendations, are presented below. The Summary's findings and recommendations are not presented in priority order but highlight conclusions reached in the report. As a result, not all of the report's findings and recommendations are included in this Summary. However, the committee points out that all of the report's recommendations are important.[4]

---

[2]The committee's first report, a letter report, was delivered to the CNO in April 2010 (see Appendix D). The present report, the committee's final report, accords with the findings and recommendations in the committee's letter report and provides additional findings, recommendations, and analysis.

[3]For the purposes of this report, in making recommendations for naval leadership actions, the term "immediate" is defined as requiring action now through the next Program Objective Memorandum (POM) cycle, in this case POM-14; "near term" as requiring close monitoring with action anticipated to be needed within the next 10 years; and "long term" as requiring monitoring with action anticipated to be needed within 10 to 20 years.

[4]Based on the judgment of this committee and the best available data, and to help provide a more quantitative assessment for a range of uncertain possible outcomes, an outcome termed "likely" has at least a two-thirds chance of occurring, and an outcome termed "very likely" has at least a 90 percent chance.

SIX AREAS FOR U.S. NAVAL LEADERSHIP ACTION

Action Area 1: Support ratification of the United Nations Convention on the Law of the Sea.

In May 2010, the CNO stated that the need for U.S. formal participation in the United Nations Convention on the Law of the Sea (UNCLOS) becomes more pressing as ice continues to melt in the Arctic.[5] Other U.S. naval leaders have expressed similar views.[6] The geopolitical situation in the Arctic region has become complex and nuanced, despite the area being essentially ignored since the end of the Cold War. The Arctic Council, a governmental forum of the five Arctic nations (Canada, Denmark, Norway, Russia, and the United States) plus Iceland, Sweden, and Finland, offers a diplomatic vehicle for addressing contemporary Arctic issues. However, maritime boundary disputes abound. For example, Canada and the United States, and Canada and Denmark have unresolved territorial sea and exclusive economic zone disputes in the Arctic. Norway and Russia disagree over offshore areas around Svalbard. The status of the Northwest Passage through the Canadian archipelago—internal Canadian waters or an international

[5]The committee studied the implications of the failure of the United States to ratify the 1982 UN Convention on the Law of the Sea (UNCLOS) from the standpoint of potential impacts on national security due to climate change. In this regard, the committee's perspectives are in line with those of Department of Defense (DOD) leadership, including the Secretary of Defense, the Chairman of the Joint Chiefs of Staff, the Secretary of the Navy, the Chief of Naval Operations, the Commandant of the Marine Corps, and the Commandant of the Coast Guard regarding ratification of UNCLOS. For example, the 2010 DOD Quadrennial Defense Review provides endorsement for U.S. ratification of UNCLOS in its discussion of climate and energy (see *Quadrennial Defense Review,* February 2010, p. 86 [p. 108 of the PDF file], available at http://www.nationaljournal.com/congressdaily/issues/graphics/Defense-Review-2010.PDF). The committee realizes that the U.S. ratification of UNCLOS involves a number of nonmilitary issues. For additional reading, see Ronald O'Rourke, 2010, *Changes in the Arctic: Background and Issues for Congress*, March 30, Congressional Research Service, Washington, D.C., pp. 6-7; and National Intelligence Council, 1996, *Law of the Sea, The End Game, Intelligence Community Assessment,* March. Available at http://www.dni.gov/nic/special_endgame.html. A previous National Research Council committee, also operating under the auspices of the Naval Studies Board, examined UNCLOS and the international legal framework in the context of maritime security partnerships. See National Research Council, 2008, *Maritime Security Partnerships,* The National Academies Press, Washington, D.C.

[6]For example, in a May 2010 speech at the National Press Club, ADM Gary Roughead, Chief of Naval Operations, strongly endorsed the United Nations Convention on the Law of the Sea as "the vehicle by which we can collectively provide continuing stability in the maritime domain" (see *Inside Defense,* 2010. "Roughead Goes to Bat for Ratification of the Law of the Sea Treaty," May 24). ADM Thad Allen, former Commandant, U.S. Coast Guard, has also issued public statements supporting ratification of UNCLOS: in 2009, he provided testimony on UNCLOS to the United States Senate Appropriations Subcommittee on Homeland Security (see http://www.jag.navy.mil/organization/code_10_law_of_the_sea.htm, and http://www.hstoday.us/content/view/9912/149/). Also, ADM James G. Stavridis, USN, Commander of the United States European Command, and Supreme Allied Commander, Europe, stated in a February 2010 meeting with the committee that the United States should ratify UNCLOS.

strait—has been a Canadian concern since at least 1985. The issue is not resolved, and current transits are allowed through nation-to-nation bilateral agreement for icebreaker transits.

The most notable maritime boundary issues involve existing and potential claims of the extended outer continental shelf under provisions of the UNCLOS. National Security Presidential Directive-66, which inter alia outlines national security interests of the United States in the Arctic, raised the possibility that Arctic issues will require national security attention in the future.

FINDING: *The committee has studied the implications of the failure of the United States to ratify the 1982 United Nations Convention on the Law of the Sea (UNCLOS) from the standpoint of potential impacts on national security in the context of a changing climate. As climate change affords increased access to the Arctic, it is envisioned that there will be new opportunities for natural resource exploration and recovery, as well as increased ship traffic of all kinds, and with that a need for broadened naval partnership and cooperation, and a framework for settling potential disputes and conflicts. By remaining outside the Convention, the United States makes it more difficult for U.S. naval forces to have maximum operating flexibility in the Arctic and complicates negotiations with maritime partners for coordinated search and rescue operations in the region. (Chapter 1)*

RECOMMENDATION: *The ability of U.S. naval forces to carry out their missions would be assisted if the United States were to ratify UNCLOS. Therefore, the committee recommends that the Chief of Naval Operations, the Commandant of the Marine Corps, and the Commandant of the Coast Guard continue to put forward the naval forces' view of the potential value and operational impact of UNCLOS ratification on U.S. naval operations, especially in the Arctic region. (Chapter 1)*

Action Area 2: Prepare for increased strain on capabilities due to greater humanitarian assistance/disaster relief (HA/DR)-related missions, as well as the opening of new international and territorial waters in the Arctic.

Greater HA/DR-Related Missions

Numerous peer-reviewed reports and scientific models anticipate a range of increasing global stresses due to the effects of climate change alone and in combination with other environmental stressors, such as projected global population growth. These reports and models suggest more severe or frequent droughts, floods, storms, and other events with negative consequences for food and water supplies, possibly leading to even greater stress on the expanded human population.

From a national security standpoint, such climate change effects would likely amplify stresses on weaker nations and generate geopolitical instability in already vulnerable regions. Furthermore, naval missions may be impacted from such effects, including the sorts of antipiracy and counterterrorism missions now being conducted off Somalia. However, the greatest impact to naval missions will be an increase on HA/DR-related missions. In short, these additional HA/DR-related missions resulting from projected climate change will have the potential to strain military resources and existing national security missions.

The U.S. Navy, as a forward-deployed force, is in position to reach disaster relief sites faster than other agencies and will almost assuredly experience increased demand for assistance if disasters increase due to climate change. The demand for Naval Construction Force capability and Navy hospital ships in support of HA/DR-related missions is likely to increase in proportion to the operational tempo of U.S.-sponsored international HA/DR operations. Likewise, the U.S. Marine Corps, with its forward-deployed Marine Expeditionary Units (MEUs), should expect to be called upon to assist with extreme weather-related HA/DR missions. However, the pace and extent of this increase are as yet unknown.

FINDING: *The unique capability provided by the U.S. Navy hospital ships will become even more important in supporting potential humanitarian assistance/disaster relief (HA/DR)-related missions that will likely occur as a result of crises created by climate change. The Navy needs to maintain this capability beyond the life of its current two-ship hospital fleet. (Chapter 2)*

RECOMMENDATION: *The Program Executive Office for Ships (PEO-Ships), the Naval Sea Systems Command (NAVSEA), and the Military Sealift Command (MSC) should analyze alternatives to retain the medical capability of the current hospital ships into the future. The analysis should address construction of new military or commercial platforms like the Mobile Landing Platform (MLP) that will join the Maritime Prepositioning Force (MPF); modification to current surface platforms or amphibious "big-decks"; or construction of next-generation Navy fleet hospitals to meet the requirements. In this context, PEO-Ships, NAVSEA, and MSC should also explore the feasibility of leasing commercial ships and crews to meet the requirements, but in doing so must ensure that the provisions for operating rooms, sophisticated trauma care, and guaranteed availability on very short notice are included. (Chapter 2)*

FINDING: *Global climate change projections from the Intergovernmental Panel on Climate Change* Fourth Assessment Report *(AR4) suggest damaging impacts in developing and developed nations that may be destabilizing in many parts of the world. These projections would affect U.S. national security and stress naval resources. In particular, naval forces will likely be required to carry out more* *frequent humanitarian assistance/disaster relief (HA/DR)-related missions. At the same time, U.S. naval forces would be expected to execute their ongoing national security military missions and to position themselves for supporting missions in destabilized regions around the globe. It is also expected that the demand for U.S. Naval Construction Force and Marine Expeditionary Unit capabilities will increase in proportion to the operational tempo of U.S.-sponsored international HA/DR missions. (Chapter 2)*

RECOMMENDATION: *In the near term, the Chief of Naval Operations (CNO) should not specifically fund new force-structure capabilities to deal with the effects of projected climate change; however, the CNO should begin to hedge against climate change impacts through planning for modifications of the existing force structure as climate change requirements become clearer. The U.S. naval forces (the U.S. Navy, Marine Corps, and Coast Guard) should begin to consider potential specific force-structure capabilities and training standards for conducting missions arising from, or affected by, climate change, particularly HA/DR-related missions. (Chapter 2)*

Opening of New International and Territorial Waters in the Arctic

U.S. Navy, Marine Corps, and Coast Guard leaders have recognized the potential impact of projected climate change on naval operations and capabilities, especially with respect to the Arctic. In May 2009, the CNO established Navy Task Force Climate Change (TFCC); in November 2009, TFCC issued its first report—U.S. Navy Arctic Roadmap—that offered a chronological listing of Navy action items, objectives, and desired effects to address climate-change-related Arctic issues for FY 2010 to FY 2014.[7]

As this report makes evident, the committee fully agrees with TFCC's initial report. Indeed, recent climate change may have the most immediate and obvious implications for maritime operations in the Arctic region. The Arctic is experiencing dramatic effects due to recent trends in global climate, including significant reductions in sea-ice cover in the Arctic Ocean and the disappearance of older, thicker, multiyear ice. This committee expects the decline in Arctic summer sea ice to continue at the current rate (10 percent per decade) or more in the next few decades. This would allow "ice-free" access over large stretches of the Arctic in late summer by 2030 that would be sufficient for reliable cross-Arctic transit.[8]

As a result of reduced multiyear ice, the Arctic Ocean is rapidly acquiring the

[7]The U.S. Navy Arctic Roadmap is available at http://www.navy.mil/navydata/documents/USN_artic_roadmap.pdf.

[8]Throughout this report, the term "ice-free" is used to mean that multiyear ice has nearly (or completely) disappeared; however, to date, in what are termed "ice-free" conditions, sufficient ice is present to remain a hazard to ordinary ships and routine marine operations.

types of maritime activities in the summer months that normally occur elsewhere in the world's ice-free oceans. Related to the increasing accessibility of Arctic waters, the U.S. Geological Survey has reported that significant natural resources (oil, natural gas, and nonfuel minerals) may become available for exploitation as ice melts and climate tempers. Recent studies have also shown that despite the continued harsh conditions, maritime tourism is expanding in the Arctic, especially around Greenland and Svalbard but also in the Northwest Passage and around Arctic Alaska.

However, the Navy currently has limited surface capability in the Arctic, and its supporting operational infrastructure in the region is severely limited versus the growing security demands in this increasingly accessible maritime domain.

FINDING: *The nation has very limited icebreaker capability, which could limit the U.S. ability to train, operate, and engage in the Arctic. Furthermore, as noted in a 2007 National Research Council report, "both operations and maintenance of [the] polar icebreaker fleet have been underfunded for many years, and the capabilities of the nation's icebreaking fleet have diminished substantially" and, among other things, "the U.S. Coast Guard [USCG] should be provided sufficient operations and maintenance budget[s] to support an increased, regular, and influential presence in the Arctic."*[9] *Moreover, U.S. national icebreaker assets are old, obsolete, and under the control of another agency that does not have a national security operational mandate. The present committee believes that future USCG missions in the Arctic will require autonomy and command of their vessels. (Chapter 2)*

RECOMMENDATION: *In order to support the U.S. naval forces' missions in the Arctic, the U.S. Coast Guard (USCG) needs icebreaker capabilities under its operational control. While there are other national requirements for such ships, action should be taken to provide these operational capabilities to the USCG. Therefore, the Chief of Naval Operations should support the initiatives of the Commandant of the Coast Guard to define future USCG icebreaker needs. As such, future U.S. national icebreaker assets should be defined as part of a holistic force structure that also accommodates ongoing National Science Foundation-sponsored polar research needs. (Chapter 2)*

FINDING: *The current situation of the three combatant commanders—Commander, U.S. European Command; Commander, U.S. Northern Command; and Commander, U.S. Pacific Command—having overlapping areas of responsibility for the Arctic was perhaps workable when the Arctic was less important than it is rapidly becoming. This division of responsibility in the Arctic is inconsistent with*

---

[9]National Research Council. 2007. *Polar Icebreakers in a Changing World: An Assessment of U.S. Needs,* The National Academies Press, Washington, D.C., p. 102.

U.S. national interests and does not match the command structure of other U.S. agencies (such as the Department of Homeland Security and the U.S. Department of State) in this increasingly significant region of the world. (Chapter 2)

RECOMMENDATION: *The Chief of Naval Operations should engage the Joint Chiefs of Staff in a review of combatant commanders' responsibilities for the Arctic, with the goal of ensuring the most effective command structure. Interagency considerations, including but not limited to the U.S. Department of State, should be included in these deliberations. (Chapter 2)*

FINDING: *In the post–Cold War era, the U.S. Navy has had a very limited surface ship presence in true northern latitude, cold-weather conditions. According to information presented to the committee, the U.S. military as a whole has lost most of its competence in cold-weather operations for high-Arctic warfare. (Chapter 2)*

RECOMMENDATION: *The Chief of Naval Operations, the Commandant of the Marine Corps, and the Commandant of the Coast Guard should establish a strong and consistently funded effort to increase Arctic operations and share lessons, including with allies. In the immediate term, the Navy should begin Arctic training and the Marine Corps should also reestablish a cold-weather training program. (Chapter 2)*

Action Area 3: Address naval coastal installation vulnerabilities due to anticipated sea-level rise and increased storm surges.

Among the many manifestations of climate change projected for the next several decades, sea-level rise is both highly certain to occur and highly certain to come with economic costs. Precision in the measurement of changes in globally averaged sea level was improved substantially in the early 1990s with the deployment of the Ocean Topography Experiment (TOPEX)/Poseidon satellite altimeter followed by later high-precision satellite altimeter missions. As a result, it is now possible to detect acceleration in sea-level rise over the past few decades. The current estimated rate (3 mm/year) is already at the upper limit of the range of global sea-level rise projections that were presented 20 years ago in the first IPCC assessment. Although this rate is small relative to the magnitude of tidal excursions at most localities, the probability of the sea level rising at this rate or faster (while adding to tidal excursion and storm surges) over the next century requires serious assessment of the implications for coastal facilities.

Although a great deal of attention has focused on the question of mean sea-level rise, it is the regional variations that are of most serious concern to naval forces and their installations. Worst-case regional changes are more than an order of magnitude greater than the global mean. In many situations, neither regional nor global sea level is directly of primary interest. Rather it is the increased vulnerability to extreme events (storm surges) and their dependence upon changes in regional relative sea level, tidal amplitudes, and the nature of extreme meteorological forces that are of greatest importance. Evaluating future risks involves an understanding of changes in storm frequency and intensity as well as local sea-level rise. Each naval facility has a unique configuration and requires ongoing evaluation of changing risks as the climate changes.

FINDING: *Peer-reviewed literature since the 2007 Intergovernmental Panel on Climate Change* Fourth Assessment Report *(AR4) suggests that loss of ice from small ice bodies (e.g., mountain glaciers and small ice caps) may have been underestimated in the last IPCC report and that major changes in Greenland and Antarctic ice sheet dynamics can take place over relatively short timescales. Sea-level variations caused by shifts in wind, rain, evaporation, and land-ice volume can cause far greater local changes in sea-level variations than the global mean rise that is projected from thermal expansion of the ocean and land-surface meltwater runoff. (Chapter 3)*

RECOMMENDATION: *Based on recent peer-reviewed scientific literature, the Department of the Navy should expect roughly 0.4 to 2 meters global average sea-level rise by 2100, with a most likely value of about 0.8 meter. Projections of local sea-level rise could be much larger and should be taken into account for naval planning purposes. However, U.S. naval leadership (e.g., the Oceanographer of the Navy) should be aware that this estimate is subject to change, and it should be reviewed routinely for any significant change. (Chapter 3)*

FINDING: *Neither regional nor global sea level is of primary interest in determining naval coastal installation vulnerability. Rather, it is the increased vulnerability associated with extreme events (storm surges) and their dependence on changes in regional sea level, tidal amplitudes, and the nature of extraordinary meteorological forces that are of greatest importance. (Chapter 3)*

FINDING: *U.S. Navy, Coast Guard, and Marine Corps coastal installations around the globe will become increasingly susceptible to projected climate change. Several assessments now under way on naval installation vulnerabilities appear to be focused primarily on static sea-level rise and coastal inundation only. According to these current assessments, some adaptive actions are indicated owing to already identified vulnerabilities at specific naval installations. The preliminary review of climate-change-related base vulnerabilities across the DOD—currently under way as directed by the 2010 Quadrennial Defense Review[10]—does not include some important factors that affect coastal installation*

[10]Secretary of Defense (Robert M. Gates). 2010. *Quadrennial Defense Review,* Department of Defense, Washington, D.C., February.

vulnerabilities, although it provides a baseline assessment across all branches of the armed services and serves as a starting point for more in-depth analysis and action. (Chapter 3)

RECOMMENDATION: *The Commander, Naval Installations Command, and the Navy Director for Fleet Readiness and Logistics should work with their U.S. Coast Guard and Marine Corps counterparts—and in conjunction with the other armed services and the Office of the Secretary of Defense—to ensure that a coordinated analysis is undertaken to address naval-installation vulnerability to rising sea levels, higher storm surges, and other consequences of climate change. In performing this vulnerability analysis, naval facility managers should recognize that each and every naval facility has a unique configuration and requires ongoing oversight of the changing risks as the climate system shifts. For example, local storm surge impact in climate-induced extreme storm events is likely to represent a bigger vulnerability than sea-level rise alone. (Chapter 3)*

RECOMMENDATION: *For Program Objective Memorandum (POM)-14 planning purposes, the Chief of Naval Operations should prepare to invest in early-stage adaptation for targeted low-elevation naval installations identified in current vulnerability assessments as being at "very high risk" from more intense storm surges, sea-level rise, and other climate change impacts. Other risks for naval installations as a result of projected climate change require further analysis and planning at this time, but no immediate direct additional substantial investment beyond current budget plans. (Chapter 3)*

Action Area 4: Address U.S., allied, and/or international maritime partnership demands based on climate change scenarios.

U.S. allies and their militaries will face national security challenges similar to those faced by the United States and its naval forces as a result of climate change. As climate change influences the geopolitical landscape, demands are expected to increase for HA/DR and maritime security missions and, in some cases, potential Arctic engagement. However, internal economic and political pressure, as well as geographical proximity to climate-change-influenced geopolitical hot spots, will lead to different responses from U.S. allies and their partners. Some allies will have an inherently greater capacity than others, and some may be required to deal with severe local climate-change-related issues internally or just across their borders.[11] According to information presented to the committee, several climate-

[11]For example, the 2008 report *Global Trends 2025* by the National Intelligence Council and the World Bank *World Development Report 2010: Development and Climate Change* suggest that given the ecological and socioeconomic characteristics in Northern Africa and major segments of Sub-Saharan Africa, and the current trend of water stress and desertification in those regions, the human

change-related global hot spots will be of particular concern to the United States and its allies.

Based on these geopolitical hot spots, projected climate change will affect U.S. allies in varying ways domestically and regionally. While these challenges are unlikely to trigger any treaty obligations (under NATO; the Australia, New Zealand, United States Security Treaty; or the U.S.-Japan Security Treaty, for example), it is likely that allies may request U.S. assistance, particularly in dealing with humanitarian assistance, disaster relief, and mass migration. Traditionally, the posture of the United States has been to assist allies to the greatest extent possible.

The historical record of U.S. military support for global HA/DR-related missions suggests that the President of the United States is likely to continue directing U.S. naval forces to respond to hot spots around the globe as a result of climate change contingencies. The capabilities and willingness of U.S. allies and their partners to participate in these responses will be critical because the United States will lack the resources and, in some instances, the strategic justification for responding alone to every request for assistance in dealing with climate-related contingencies, even when U.S. interests may be directly at stake. More robust partnerships will be required to deal with climate-change-related issues.

Given the scope and scale of potential climate change contingencies, and the projected global climate change vulnerabilities, the United States and its naval forces will want to cooperate with allies, non-allies, and private organizations in both anticipating and responding to global climate change and geographic hot spots. At this time, these partnerships either are not sufficiently robust or are not tailored for the quantity and type of missions that are most likely to occur, including the need for additional partnerships for the United States to properly deal with Arctic issues.

FINDING: *All regions of the world will experience the effects of projected climate change. Some climate change effects, such as changes in storm patterns and drought, will have direct impacts in the United States. Should regional storms and droughts intensify over time they may well drive mass migrations to the United States from neighboring countries, including Mexico, the Caribbean, and Central America. Projected climate change will also directly and indirectly affect most U.S. allies, including NATO countries, Australia, Japan, and all other major non-NATO allies, which in turn may request or require U.S. assistance. (Chapter 4)*

---

impact of climate change may be more marked there than in other regions of the world, potentially leading to mass migration pressures that would also impact Southern Europe. See The World Bank, 2009, *World Development Report 2010: Development and Climate Change,* November, The World Bank, Washington, D.C.; and National Intelligence Council, 2008, *Global Trends 2025: A Transformed World,* November, p. 53; available at http://www.dni.gov/nic/PDF_2025/2025_Global _Trends_ Final_Report.pdf. Accessed May 25, 2010.

RECOMMENDATION: *Given that U.S. naval forces cannot be fully prepared for or respond to all plausible climate contingencies, the Chief of Naval Operations, working with the combatant commanders, the Commandant of the Coast Guard, and the Commandant of the Marine Corps, should develop or expand maritime partnerships with other nations. Projected climate change will affect all regions of the world, and so U.S. naval forces should seek to develop these partnerships with long-standing allies and nontraditional partners alike, including Russia, China, and nongovernmental organizations. In particular, developing climate change response capabilities within the NATO alliance could strengthen global climate change response capabilities and the alliance itself. (Chapter 4)*

FINDING: *Although the likelihood of conflict in the Arctic is low, it cannot be ruled out, and competition in the region is a given. However, cooperation in the region should not be considered a given, even with close allies. Although there are mechanisms for bilateral and multilateral cooperation in the area, including the Arctic Council, these relationships and mechanisms are largely untested for emerging conditions. Additionally, with the ratification of UNCLOS, U.S. naval forces will be better positioned to conduct future naval operations and protect national security interests, especially in the Arctic. (Chapter 4)*

RECOMMENDATION: *The Chief of Naval Operations, working with the combatant commanders, the Commandant of the Coast Guard, and the Commandant of the Marine Corps, should build maritime partnerships in the Arctic region and encourage the United States to continue to identify and adopt policies and relationships in the Arctic that will build cooperation for new circumstances and minimize the risks of confrontation. (For example, naval leaders should pursue bilateral and multilateral training and exercising of U.S. naval personnel with partner nation personnel in maritime security, search and rescue, and HA/DR, and continue strong support of the U.S. efforts in the Arctic Council.) There should be no assumption that the geostrategic situation will take care of itself or that U.S. interests in the region are currently protected and promoted. (Chapter 4)*

Action Area 5: Address the potential impacts on the technical underpinnings that enable, in part, naval force capabilities, especially any impacts due to the necessity to operate in polar regions.

The technical underpinnings that enable, in part, naval forces' capabilities are sophisticated, widely available, and reliable throughout the temperate and tropical oceans, and they are therefore often taken for granted. While the effects of climate change do not directly affect these underpinnings, they mandate that naval forces operate in areas that present challenges for supporting systems and infrastructure and, ultimately, challenges to overall capabilities. Indeed, there is a high likelihood that a warming climate will increase the operational tempo in polar regions and consequently intensify the demands on navigation systems, communication systems, and nautical charts. The initial increase in tempo will be driven by scientific and exploratory missions, especially so in the Arctic. However, navigation in the polar regions is challenging not only due to sea-ice and adverse weather conditions but also due to limitations of current navigation systems and technologies at high latitude which are degraded relative to the performance in other regions of the world.

On the other hand, there are no significant first-order effects from climate change on U.S. antisubmarine warfare (ASW) capabilities. A robust infrastructure that collects, analyzes, and distributes oceanographic data essential to ASW effectiveness is in place and covers active submarine operating areas adequately. Climate change will, however, mandate that submarine and ASW operations become more robust in the Arctic Ocean, where essential data are sparse or nonexistent in both spatial and temporal senses. Moreover, as potential adversarial submarines have become acoustically more quiet, ASW operations have evolved away from a pure submarine-on-submarine mission to a cooperative, coordinated mission involving fixed and mobile sensors and surface, subsurface, and air platforms. This extensive and deployable ASW infrastructure that supports the principal nuclear-powered attack submarine (SSN) hunter platforms is generally deployed in the temperate oceans but would be challenged to operate in the Arctic and does not presently do so. As well, the supporting tactical oceanographic data collection, analysis, and distribution system does not extend to the Arctic, although it must be established or restored to enable effective ASW operations in that region, which will become an inevitable national imperative.

FINDING: *U.S. military navigation and communications systems have been optimized to support operations in non-polar regions. Likewise, data on terrain elevation and bathymetry to support military operations and nautical charting are of low resolution and sparse in the Arctic. Moreover, while accurate ice coverage charts are available to guide surface navigation, reliable real-time ice characterization and maps in emergent Arctic transit routes are not. The combined effect of degraded navigation, communications, and charting systems could impact safe operations and reduce the performance of military systems in the polar regions. (Chapter 5)*

RECOMMENDATION: *The Assistant Secretary of the Navy for Research, Development, and Acquisition should increase research and development efforts at the Office of Naval Research and the Naval Research Laboratory to address the operational shortfalls of existing and planned navigation, communications, and charting systems, leveraging both local and global augmentation technologies. In conjunction with the National Oceanic and Atmospheric Administration, the Department of the Navy should increase priority for extending modern navigation, communications, and charting coverage to include the Arctic region. (Chapter 5)*

FINDING: *The United States had an Arctic research program during the Cold War that has essentially ceased. Moreover, there is no infrastructure to support antisubmarine warfare (ASW) in the Arctic. While there are no significant ASW activities now in the Arctic, U.S. naval forces need to be prepared to operate there safely. The United States' diminished Arctic research program and capabilities from what existed during the Cold War—plus the need for even better performance from its ASW systems—put U.S. naval forces' ability to operate as needed in the Arctic at risk if the United States does not keep pace with the capabilities of other Arctic nations, especially Russia with its extensive claims of Arctic sovereignty, as well as with non-Arctic nations, such as China. (Chapter 5)*

RECOMMENDATION: *Given that climate change may drive the U.S. naval forces to conduct antisubmarine warfare (ASW) operations in the Arctic, the Department of the Navy should increase its submarine Arctic presence for training purposes, extend its supporting ASW oceanographic data infrastructure to the Arctic Ocean, and begin to conduct multiplatform ASW training exercises in the Arctic. Specifically, this should include:*

- *Increased research for Arctic passive and active sonars;*
- *Long-range planning to install facilities that support Arctic ASW, such as refurbishing and expanding the fixed array systems;*
- *Planning for aircraft support from the new P8;*
- *Development of high-latitude communications systems for relaying tactical and environmental data;*
- *Identifying ports for emergencies; and*
- *Incorporation of a more robust under-ice capability on Virginia-class submarines. (Chapter 5)*

Action Area 6: Support investments for additional research and development that have implications for future naval force operations and capabilities and might not be met by other groups pursuing climate-related research.

Naval operations and capabilities require, in part, knowledge of environmental information in the form of observations, model-based analysis products, and model forecasts for navigation, communication, general fleet support, ASW, and search and rescue. There are also many ocean, atmosphere, cryosphere, and land measurements needed by the Navy. Currently, global measurements in the marine environment come from a mix of Earth observing satellites and in situ sensors as part of the Global Ocean Observing System (GOOS) and the Global Climate Observing System (GCOS). There are also Department of the Navy (DoN) and Department of Defense (DOD) measurement assets geared toward addressing the needs of their specific mission sets but which contribute to the ocean observing system. In addition, classified DoN and DOD measurement assets could make significant contributions to GOOS if more open access were provided. As one example, release of images of Arctic sea ice from 1999 to the present as part of the Measurements of Earth Data for Environmental Analysis (MEDEA) Program is providing unique and fundamentally new information on the loss of Arctic sea ice that is largely attributable to climate change.[12] The early MEDEA Program resulted in increased U.S. Navy collaboration and cooperative experiments with non-U.S. Navy entities, and this could serve as a model for today's efforts to expand maritime operational data and knowledge in the Arctic. In a related example, the use of U.S. Navy submarines in the interagency Science Ice Exercise (SCICEX) Program has provided unique three-dimensional under-ice oceanographic data, including very valuable upward looking sonar ice-draft measurements.

As the Navy considers the use of these measurements and the potential impact of climate change on its operations and capabilities, it is clear that evolutionary and transformational advances may be required to improve modeling and prediction of seasonal, decadal, and beyond (century-scale) climate change. At the same time, it is quite challenging to assess climate model value or success because simulations from even the most advanced modeling systems have considerable spread and uncertainty. Included in this uncertainty is the possibility of unexpected rapid changes, extreme events, or abrupt climate change associated with potential fast processes either not resolved or resulting from unaccounted-for interactions/feedbacks among different Earth system components. While progress has been made to improve climate models, there is no capability for coupled ocean-atmosphere-land-cryosphere modeling in the Navy, and there are no programs focused on seasonal-to-decadal timescale prediction to support strategic decisions related to operations, platforms, and facilities. Because of its presence on the global oceans, its long-term global ocean/ice observations and data collection, and its unique physical assets, the U.S. Navy can both benefit from and contribute strongly to a better understanding of the ocean component of climate science.

In an emerging area, one aspect of ocean acidification that might be of special importance to the Navy—the potential effects of a pH decrease on sound absorption—is still under study. The Navy should continue to monitor the research in ocean acidification closely, as the results may have potentially important implications for ocean acoustics critical to U.S. naval operations.

FINDING: *Open access to previously classified Navy data and to other Department of Defense assets through the MEDEA Program has enabled advances in climate change research that have benefited the scientific community studying climate change. A clear example of this benefit is the analysis of submarine upward*

[12]National Research Council. 2009. *Scientific Value of Arctic Sea Ice Imagery Derived Products,* The National Academies Press, Washington, D.C.

looking sonar, which shows that sea ice has been thinning in response to climate change. (Chapter 6)

RECOMMENDATION: *The Chief of Naval Research, the Oceanographer of the Navy, and the Commander, Naval Meteorology and Oceanography Command, should consider findings by the MEDEA Program (and take lessons from MEDEA actions within the intelligence community) to develop and support a Navy philosophy for providing access to previously classified information that can be used by the climate research community. Such actions would enhance the potential of these researchers to help the Navy better prepare for its mission in a future with a warmer climate. (Chapter 6)*

FINDING: *The Navy has billions of dollars in assets exposed to the threats of climate change, and it must make strategic decisions in the face of considerable uncertainty about the pace, magnitude, and regional manifestations of climate change. Yet Navy research at present has no capability for modeling the coupled ocean-atmosphere-land-cryosphere system and how it will respond to greenhouse gas forcing. The Navy also has no programs in seasonal-to-decadal timescale climate forecasting to help guide long-range strategic planning for operations, platforms, and facilities; it relies almost entirely on civilian agencies and international assessments to inform its policies and practices related to climate change. (Chapter 6)*

RECOMMENDATION: *The Assistant Secretary of the Navy for Research, Development, and Acquisition (ASN RDA) should examine the U.S. Navy's overall research and development capabilities vis-à-vis climate studies, especially with respect to coupled models and climate forecasting on seasonal-to-decadal timescales. The ASN RDA should give special emphasis to regional aspects of sea-level rise, and sea-ice concentration and extent, because of their relevance to coastal infrastructure and operational needs. The Department of the Navy should also become actively engaged in the development of an Arctic Observing System, specifically with respect to development and deployment of in situ and remote sensing systems (i.e., gliders, buoys, and satellites) as well as icebreakers in support of research. (Chapter 6)*

1

Introduction

> A series of powerful cross-cutting trends, made more complex by the ongoing economic crisis, threatens to complicate international relations and make the exercise of U.S. statecraft more difficult. The rising demand for resources, rapid urbanization of littoral regions, the effects of climate change, the emergence of new strains of disease, and profound cultural and demographic tensions in several regions are just some of the trends whose complex interplay may spark or exacerbate future conflicts.
>
> —U.S. Department of Defense, 2010, *Quadrennial Defense Review*

The February 2010 *Quadrennial Defense Review* notes that climate change will play a significant role in the future security environment for the United States.[1] Concurrently, the United States Department of Defense (DOD) and its military services are already developing policies and plans to understand and manage the effects of climate change on military operating environments, missions, and facilities. For the Navy, the Chief of Naval Operations (CNO) established Navy Task Force Climate Change (TFCC) that was charged initially with developing a road map for Navy actions in the Arctic, and then with addressing long-term Navy policy, strategy, and plans as a result of climate change.[2] This National Research Council study, commissioned by the CNO and convened under the auspices of the Naval Studies Board (NSB), was tasked to provide an understanding of the national security implications of climate change for U.S. naval forces. The study's terms of reference charge the committee to produce two reports over a 15-month period.[3] The terms of reference direct that the committee in its two reports do the following:

1. Examine the potential impact on U.S. future naval operations and capabilities as a result of climate change. . . .
2. Assess the robustness of the Department of Defense's infrastructure for sup-

[1]Secretary of Defense (Robert M. Gates). 2010. *Quadrennial Defense Review,* Department of Defense, Washington, D.C., February.

[2]See Vice Chief of Naval Operations (ADM Jonathan W. Greenert, USN). Memorandum 4000 Ser N09/9U103035, "Task Force Climate Change Charter," October 30, 2009.

[3]The study's terms of reference are provided in Appendix A. The terms of reference were formulated by the Chief of Naval Operations (CNO) staff in consultation with the NSB chair and director.

porting U.S. future naval operations and capabilities in the context of potential climate change impacts. . . .

3. Determine the potential impact climate change will have on allied force operations and capabilities. . . .
4. Examine the potential impact on U.S. future naval antisubmarine warfare operations and capabilities in the world's oceans as a result of climate change; specifically, the technical underpinnings for projecting U.S. undersea dominance in light of the changing physical properties.

The committee's first report, a letter report delivered to the CNO in April 2010, summarized near-term challenges and provided findings and recommendations for U.S. naval forces to address the more immediate climate-change-related challenges and planning issues.[4] This report represents the committee's final report and provides a more complete examination of issues identified in the study's terms of reference.

ASSUMPTIONS FOR THE REPORT

The study's terms of reference direct that this study be based on Intergovernmental Panel on Climate Change (IPCC) assessments and other subsequent relevant literature reviewed by the committee. Therefore, the committee did not address the science of climate change or challenge the assessments on which the committee's findings and recommendations are based. This report addresses both the immediate and the long-term climate-change-related challenges for U.S. naval forces for each of the four areas of the terms of reference and provides findings and recommendations for addressing these challenges.[5] Additionally, this report identifies research and development needs for U.S. naval forces within the context of a changing climate, and it provides findings and recommendations that the committee believes will assist in reducing underlying uncertainties for naval planning and missions.

This chapter begins with an overview of climate change effects and their implications for national security. It then examines increased international activity in the Arctic as a result of climate change and the resulting implications of the United Nations Convention on the Law of the Sea (UNCLOS) for U.S. naval forces. Following this, the chapter reviews the positioning of the U.S. naval leadership on climate change and provides a summary of additional relevant climate assessments. The chapter concludes with a discussion of risk management

[4]The committee's letter report is provided in Appendix D.

[5]For the purposes of this report, in making recommendations for naval leadership actions, the term "immediate" is defined as requiring action now through the next Program Objective Memorandum (POM) cycle, in this case POM-14; "near term" as requiring close monitoring with action anticipated to be needed within the next 10 years; and "long term" as requiring monitoring with action anticipated to be needed within 10 to 20 years.

approaches for addressing future climate uncertainties and an overview of the remainder of the report.

CLIMATE CHANGE EFFECTS

There is broad scientific consensus on many climate change topics.[6] These climate certainties include measured or observed (1) higher surface, troposphere, and ocean temperatures; (2) more precipitation and drought extremes; (3) melting of mountain glaciers, Arctic sea ice, and ice sheets; and (4) a rising sea level.[7] Each of the four certainties has the potential to impact U.S. naval forces' operations and installations; if continued as projected, many will have national security implications as well. In most cases, the effects of climate change can be summarized through the effects on water—prolonged droughts, more intense storms and floods, melting ice, and/or changing ocean conditions, including ocean acidification. Many of these climate change effects and their resulting impacts are summarized in Box 1.1 and illustrated in Figures 1.1 and 1.2. In many regions of the world, the impact of climate change is likely to further exacerbate the preexisting stress on water supplies and the associated mounting pressures of population growth.[8] While these issues and other potential climate change impacts are important, this report is focused through the lens of relevance and implications for U.S. naval forces.

---

[6]For example, the National Research Council of the National Academies has recently released the first four reports from a congressionally mandated suite of studies known as America's Climate Choices. These studies are discussed later in this chapter, and additional information is provided at the America's Climate Choices website: http://americasclimatechoices.org/, accessed July 28, 2010.

[7]The U.S. Global Change Research Program, composed of 13 federal agencies, reported in 2009 that climate-related changes are already being observed in every region of the world, including the United States and its coastal waters. Among these physical changes are increases in heavy downpours, rising temperature and sea level, rapidly retreating glaciers, thawing permafrost, lengthening growing seasons, lengthening ice-free seasons in the oceans and on lakes and rivers, earlier snowmelt, and alterations in river flows. See Thomas R. Karl, Jerry M. Melillo, and Thomas C. Peterson, 2009, *Global Climate Change Impacts in the United States,* Cambridge University Press, New York, pp. 25-47.

[8]For example, Columbia University's Center for International Earth Science Information Network (CIESIN) has compiled information from IPCC assessments, the 2005 World Bank report *Natural Disaster Hotspots: A Global Risk Analysis*, and CIESIN's gridded world population data sets to present a projected geographic distribution of vulnerability in 2050. In presentations to the committee, CIESIN representatives reported that global population nearly doubled from 1968 through 2008, and that by 2048 it could grow another 40 percent, to more than 9 billion people, adding even greater stresses to water and food supplies (Robert S. Chen, Center for International Earth Science Information Network, Columbia University, "Human Dimensions of Climate Change," and Marc Levy, Center for International Earth Science Information Network, Columbia University, "Climate Change and U.S. National Security," presentations to the committee, November 19, 2009, Washington, D.C.). CIESIN also reported that population increases are fastest in areas most vulnerable to intense storms and flooding (e.g., coastal areas, islands, and river basins). The CIESIN analysis combines its population data sets with IPCC-projected climate-change-related vulnerabilities, economic data, and past disaster-related losses to identify areas at relative high risk from one or more hazards.

BOX 1.1
Measured Climate Change Effects with Impact for U.S. Naval Forces

Climate Change Effect	Observed Impact
Higher Temperature Extremes	Higher maximum temperatures, increased heat index and heat waves over land areas, harsher operating conditions; negative impacts for fresh water supply; negative impact for agricultural production; changing disease vectors (tropical diseases migrating north)
More Vigorous Hydrological Cycle	More energy in hydrological cycle, extreme rainfall events, more frequent high-intensity storms in some areas, prolonged regional droughts, flooding, potential humanitarian assistance and disaster relief impacts
Melting of Ice	Reduced glacier mass, reduced ice sheets, reduced multiyear sea ice in the Arctic, thawing permafrost, changing ocean salinity
Sea-Level Rise	Higher storm surges, salinization of fresh water, risks to coastal infrastructure, risks to high-population coastal deltas

IMPLICATIONS FOR NATIONAL SECURITY

Climate change alone is unlikely to cause conflict, but its manifestations can. The committee reviewed reports by the Center for Naval Analyses, the National Intelligence Council, and others that find that climate change can act as an accelerant of instability or conflict, placing a burden to respond on civilian institutions and militaries around the world and leading to potential national security implications.[9] In addition, extreme weather events induced by a changing

[9]See Military Advisory Board, 2007, *National Security and the Threat of Climate Change*, CNA Corporation, Alexandria, Va.; and National Intelligence Council, 2008, *2025 Global Trends Report,* November.

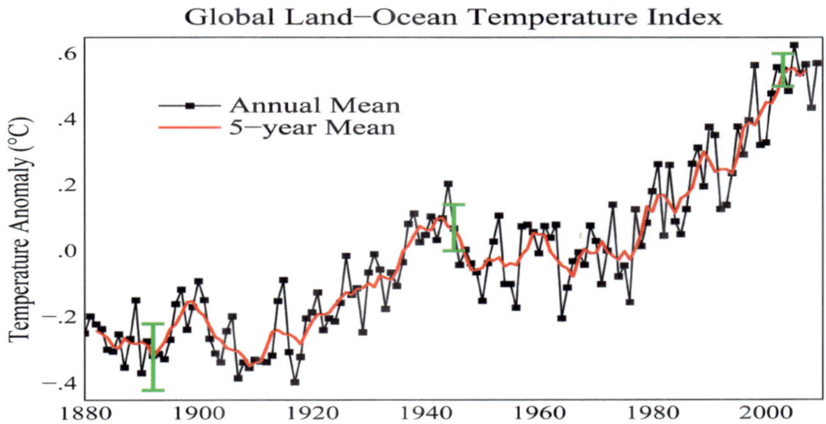

FIGURE 1.1 Climate measurements indicate that Earth is getting hotter. Except for a leveling off between the 1940s and 1970s, Earth's surface temperatures have increased since 1880. The last decade has brought the temperatures to the highest levels ever recorded. The graph shows global annual surface temperatures relative to 1951-1980 mean temperatures. As shown by the red line, long-term trends are more apparent when temperatures are averaged over a 5-year period. The hottest 14 years on record have all occurred since 1990. SOURCE: NASA/Goddard Institute for Space Studies. 2010. "Global Land-Ocean Temperature, 1880-Present," December.

climate may lead to increased demands for defense support to civil authorities for humanitarian assistance or disaster response both within the United States and globally. Viewed from a national security standpoint, these changes would likely amplify stresses on weaker nations and generate geopolitical instability in already vulnerable regions.[10]

A range of military missions may result from such conditions, including the sorts of antipiracy and counterterrorism missions now being conducted in the waters off the coast of Somalia. However, the clearest implications of these changes are for humanitarian assistance/disaster relief (HA/DR) missions, which may increase in frequency, thereby potentially straining military transportation resources and the supporting force structures. The U.S. Navy and Marine Corps, as a forward-deployed force, are in position to reach disaster relief sites faster than other agencies, and therefore will almost assuredly experience increased demand

[10]See Statement of the Record of Dr. Thomas Fingar, Deputy Director of National Intelligence for Analysis and Chairman of the National Intelligence Council, before the Permanent Select Committee on Energy Independence and Global Warming, House of Representatives, "National Intelligence Assessment on the National Security Implications of Global Climate Change to 2030," June 25, 2008. Available at http://www.dni.gov/testimonies/20080625_testimony.pdf. Accessed November 24, 2009.

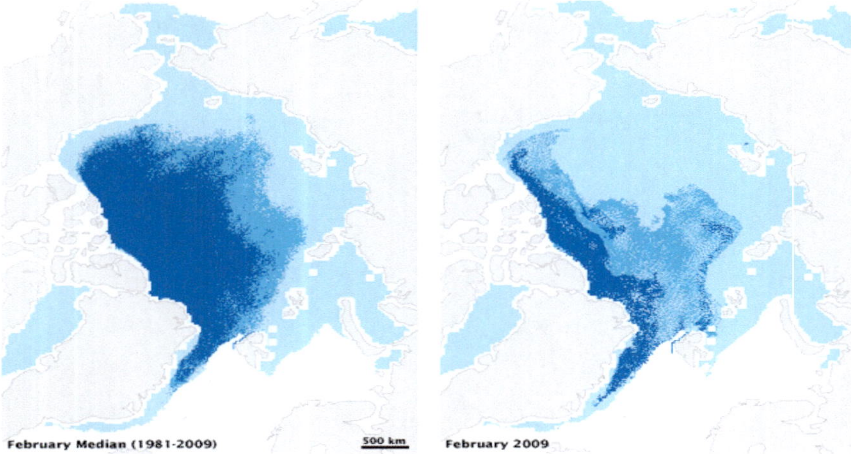

FIGURE 1.2 Climate measurements indicate that Arctic sea ice is shrinking and thinning. These Arctic maps show the median age of February sea ice, 1981–2009 (left) and February 2009 (right). As of February 2009, ice older than 2 years accounted for less than 10 percent of the ice cover. (Dark blue represents multiyear ice.) SOURCE: National Snow and Ice Data Center (NSIDC), University of Colorado, Boulder; data provided by James Maslanik and Charles Fowler, Colorado Center for Astrodynamics Research, Aerospace Engineering Sciences, University of Colorado.

for assistance if climate-related disasters increase.[11] Recent events have shown that the U.S. Navy and Marine Corps will often be called upon as first responders on behalf of the United States to provide immediate and large-scale international HA/DR assistance and to help secure U.S. interests in sensitive regions.[12] While

[11]A 2007 joint maritime strategy document for the U.S. Navy, Marine Corps, and Coast Guard calls out "humanitarian assistance and disaster response" as one of six capabilities that constitute the core of U.S. maritime power and that "reflect an increased emphasis on those activities that prevent war and build partnerships." See *Cooperative Strategy for 21st Century Seapower,* available at http://www.navy.mil/maritime/MaritimeStrategy.pdf. Accessed November 23, 2009. However, it is not the sole responsibility of the U.S. military to respond to national and international humanitarian and disaster-relief emergencies; many U.S. and international governmental and private agencies may be engaged in any given relief operation.

[12]For example, in the aftermath of Tropical Storm Ketsana striking the Philippines on September 25, 2009, the U.S. Navy and Marine Corps worked with the Philippine government (and in support of the U.S. Department of State and the U.S. Agency for International Development Office of Foreign Disaster Assistance) to rapidly provide critically needed supplies in support of disaster relief to help mitigate human suffering and prevent further loss of life. In this case, a team of approximately 100 personnel composed of Marines from the III Marine Expeditionary Force flew from Okinawa to the Philippines on September 29, 2009, to conduct humanitarian assistance assessments. On September 30, the USS *Denver*, USS *Tortuga*, and USS *Harpers Ferry*, with embarked Marines and sailors of the 31st Expeditionary Unit, set sail from Okinawa toward the Philippines. On October 1, the com-

the above scenarios are likely, the pace and extent of this increase are unknown at this time.

Of all the theaters of naval operations that the committee considered could be impacted by climate change, the Arctic[13] was found to have the most immediate challenges, all of which have relevance to each of the four areas identified in the terms of reference. For example, the reduction of Arctic sea ice has already led to increased activity in the Arctic Ocean. It is anticipated that major international maritime passages in the Arctic will be accessible by the year 2030, at least during the summer months. This change is due to the continued reductions in summer sea-ice cover in the Arctic Ocean and the rapid disappearance of older, thicker multiyear ice (see Box 1.2).[14] A result of this will be greater summer marine access and longer navigation seasons in the Arctic Ocean, both having direct implications for the U.S. Navy and Coast Guard. The January 2009 National Security Presidential Directive (NSPD)-66/Homeland Security Presidential Directive (HSPD)-25 discusses relevant U.S. national security concerns in the Arctic, including such matters as missile defense and early warning, deployment of sea and air systems for strategic sealift, strategic deterrence, maritime security operations, and ensuring freedom of navigation.[15] The U.S. Navy and Coast Guard must be positioned to meet maritime domain awareness (MDA) requirements and to recognize and respond to any potential security interests confronting the United States in this changing maritime domain.[16]

The United States Geological Survey (USGS) notes that significant natural resources (oil, natural gas, and nonfuel minerals) may become increasingly accessible for exploration and exploitation as Arctic sea ice melts on a seasonal basis. The 2008 *USGS Circum-Arctic Resource Appraisal* reports that the extensive Arctic continental shelves may constitute the geographically largest unexplored area for petroleum products remaining on Earth, with an estimated 90 billion bar-

---

manding general of the 3rd Marine Expeditionary Brigade flew from Okinawa to the Philippines to lead planning and humanitarian assistance efforts. See *U.S. Marine Corps News*. Available at http://www.okinawa.usmc.mil/public affairs/info/archive/news. Accessed November 23, 2009.

[13]In this report, the Arctic region is defined as the land and sea area north of the Arctic Circle, the circle of latitude at approximately 66.56 degrees north of the equator. The North Pole, the northernmost point of the axis around which Earth rotates, lies at the center of the Arctic region.

[14]Multiyear ice remains frozen throughout the year and is typically 2 to 5 m thick. First-year ice is formed in the winter but melts during the summer and is typically 0.3 to 2 m thick. Sea ice, as a general term, includes multiyear and first-year ice.

[15]The January 2009 National Security Presidential Directive (NSPD)-66, dual titled as Homeland Security Presidential Directive (HSPD)-25, or NSPD-66/HSPD-25, establishes the policy of the United States with respect to the Arctic region and outlines national security and homeland defense interests in the region. See National Security Presidential Directive-66, Article III B 1; available at http://www.fas.org/irp/offdocs/nspd/nspd-66.htm. Accessed July 28, 2010.

[16]U.S. naval officials define maritime domain awareness as "the effective understanding of anything associated with the maritime domain that could impact the security, safety, or economy of the United States." See *Naval Operations Concepts 2010, Implementing the Maritime Strategy;* available at http://www.navy.mil/maritime/noc/NOC2010.pdf. Accessed June 4, 2010.

BOX 1.2
A Sea-Ice Tutorial

Several forms of floating ice may be encountered by vessels at sea. The most extensive ice is that which results from the freezing of the sea surface, namely sea ice; but mariners must also be concerned with "ice of land origin"—icebergs, ice islands, bergy bits and growlers. Both icebergs and sea ice can be dangerous to shipping and always have an effect on navigation.

Young ice: Newly formed sea ice less than 30 centimeters thick is described as young ice or new ice. It forms extensively in the autumn as ocean surface temperatures fall below freezing and on leads that open in mid-winter due to shifts in the pack ice. Young or new ice is not a significant safety hazard for most Arctic vessels, although when placed under pressure by winds or currents, it can impede progress.

First-year ice: First-year ice can easily attain a thickness of 1 meter but rarely grows beyond 2 meters by the end of the winter. First-year ice is relatively soft due to inclusions of brine cells and air pockets and will not generally hole an ice-strengthened ship operated with due caution. Under pressure from winds or currents, first-year ice can impede progress to the point where even powerful vessels can become beset for hours or even days.

Old ice: If first-year ice survives the summer, it is then classified as old ice (subdivided into second-year and multiyear ice). Multiyear ice is typically 2 to 5 meters thick and is extremely hard. During the summer melt process, the brine cells and air pockets that characterize first-year ice drain out the bottom of the ice, leaving a clear, solid ice mass that is harder than concrete. Even ice-strengthened vessels are at risk of being holed by old ice. When under pressure, old ice can stop the most powerful icebreakers.

Icebergs: These are large masses of floating ice originating from glaciers. They are very hard and can cause considerable damage to a ship in a collision. Ice islands are vast tabular icebergs originating from floating ice shelves. Smaller pieces of icebergs are called bergy bits and growlers and are especially dangerous to ships because they are extremely difficult to detect.

SOURCE: *Arctic Marine Shipping Assessment 2009 Report,* Arctic Council, April 2009, p. 22.

rels of oil, 1,669 trillion cubic feet of natural gas, and 44 billion barrels of natural gas liquids still to be found in the Arctic, of which approximately 84 percent is expected to occur in offshore areas.[17] Although exploration conditions are projected to remain harsh and challenging in the Arctic, shrinking sea ice provides greater access to these potential resources.

In the committee's view, the climate-change-related changes in the Arctic hold the potential for international competition, conflict, or cooperation. The current geopolitical forces at play in the Arctic, when combined with climate model projections of continued reduction in Arctic sea ice, provide compelling evidence that future requirements for U.S. naval operations in the Arctic will significantly increase over the next 30 years.

CLIMATE CHANGE, ARCTIC CLAIMS, AND UNCLOS

Related to the increased international activity and interest in the Arctic described above, the fact that the United States has signed but not yet ratified the United Nations Convention on the Law of the Sea[18] will become even more problematic with time and as more states call for international recognition of their Arctic claims (see Box 1.3). For example, the five Arctic coastal states—Canada, Russia, Norway, Denmark (based on its territory Greenland), and the United States—are in the process of preparing Arctic territorial claims for submission to the Commission on the Limits of the Continental Shelf. Russia's claims to the Lomonosov Ridge, if accepted, would grant Russia nearly one-half of the Arctic. By remaining outside of UNCLOS, the United States seriously compromises its ability to take part in negotiations regarding the claims of other nations.[19] UNCLOS provides a legal framework for the settlement of such disputes.

The current nonparticipation of the United States in UNCLOS has serious negative implications for U.S. naval forces and their operations in the Arctic.[20] In

---

[17]See Kenneth J. Bird, Ronald R. Charpentier, Donald L. Gautier (CARA Project Chief), David W. Houseknecht, Timothy R. Klett, Janet K. Pitman, Thomas E. Moore, Christopher J. Schenk, Marilyn E. Tennyson, and Crain J. Wandrey, 2008, "Circum-Arctic Resource Appraisal: Estimates of Undiscovered Oil and Gas North of the Arctic Circle," U.S. Geological Survey Fact Sheet, 2008-3049; available at http://pubs.usgs.gov/fs/2008/3049/. Accessed June 4, 2010.

[18]An extensive discussion of the international legal framework for UNCLOS is provided in National Research Council, 2008, *Maritime Security Partnerships,* The National Academies Press, Washington, D.C., Appendix C.

[19]An overview and full text of the United Nations Convention on the Law of the Sea are available at http://www.un.org/Depts/los/convention_agreements/convention_overview_convention.htm. Accessed November 23, 2009.

[20]U.S. Navy and Coast Guard leadership have provided public testimony on the potential value and impact of UNCLOS ratification on U.S. naval operations. For example, in a May 20, 2010, speech on the Arctic at the National Press Club, ADM Gary Roughead, Chief of Naval Operations, stated that the United Nations Convention on the Law of the Sea "is the vehicle by which we can collectively provide continuing stability in the maritime domain. . . . As the only permanent member of the UN Security

BOX 1.3
Summary of the United Nations Convention on the Law of the Sea (UNCLOS) and the Arctic

UNCLOS provides an important international legal and political framework for the Arctic, with a broader aim to regulate all aspects of the resources of the sea and uses of the global ocean, including freedom of navigation. Its 320 articles and 9 annexes propose governing rules for all aspects of ocean space, including marine scientific research, commercial activities, the permissible breadth of the territorial sea (the part of the ocean nearest the shore, over which the coastal state enjoys sovereignty), maximum freedom of the seas, and the settlement of disputes relating to ocean matters.

According to UNCLOS, coastal states have undisputed sovereign rights to their territorial sea and exclusive economic zone, which extend to a distance of 200 nautical miles from their coastal baseline. Upon ratification of UNCLOS, a country has 10 years to make claims to extend its 200-nautical-mile zone. As of January 2010, over 160 parties have ratified UNCLOS. Article 76 of UNCLOS provides the rules by which coastal states may establish those outer limits. The Commission on the Limits of the Continental Shelf is the governing body for these Arctic claims.

In the case of the Arctic Ocean, the five Arctic coastal states are Canada, Denmark, Norway, Russia, and the United States. Russia ratified UNCLOS in 1997. In December 2001, Russian officials submitted a claim that 120 million hectares of underwater terrain between the Lomonosov and Mendeleev ridges be confirmed as a continuation of the Siberian shelf. Norway ratified UNCLOS in 1996 and submitted its claim in November 2006. Canada and Norway ratified UNCLOS in 2003 and 2004, respectively, and are in the process of preparing claims for submission.

The United States has not ratified UNCLOS. However, the United States is working closely with Canada to gather and analyze data through the Extended Continental Shelf Project for the submission of Canada's claim. This effort is led by the U.S. Continental Shelf Task Force, an interagency body, chaired by the Department of State with co-vice chairs from NOAA and the Department of the Interior. Both U.S. Navy and U.S. Coast Guard representatives participate on the Task Force. According to the U.S. Arctic Research Commission, the United States could lay claim to an area in the Arctic of about 450,000 square kilometers and the seabed resources therein. However, as a non-party to UNCLOS, the United States cannot participate as a member of the Commission on the Limits of the Continental Shelf; neither can the United States submit a claim under Article 76.

SOURCE: See United Nations Convention on the Law of the Sea, available at http://www.un.org/Depts/los/convention_agreements/convention_overview_convention.htm. See also http://continentalshelf.gov/.

this regard, the committee's perspectives on UNCLOS are in line with those of Department of Defense leadership, including the Secretary of Defense, the Chairman of the Joint Chiefs of Staff, the Secretary of the Navy, the Chief of Naval Operations, the Commandant of the Marine Corps, and the Commandant of the Coast Guard regarding ratification of UNCLOS.[21]

Freedom of the seas issues addressed in UNCLOS are also important for U.S. naval forces, beyond an increasingly accessible Arctic due to melting sea ice. U.S. naval forces depend upon global strategic mobility and tactical maneuverability to conduct the spectrum of sea-air-land operations in the pursuit of national interests. Similar to NSPD-66, the 2005 United States National Strategy for Maritime Security identified freedom of the seas as a top national priority.[22] Also, the Department of Defense and the Joint Chiefs of Staff discussed the major national security benefits of the Law of the Sea Convention in a 1996 report. The foremost benefit seen by this group was reported as global access to the oceans throughout the world—specifically, freedom of navigation, overflight, and telecommunications—and a stable and nearly universally accepted convention to promote public order and free access to the oceans and the airspace above them.[23]

The committee has studied the implications of the failure of the United States to ratify the 1982 UN Convention on the Law of the Sea from the standpoint of potential impacts on national security due to climate change. As climate change affords increased access to the vast Arctic, the committee envisions new opportunities for natural resource exploration and recovery as well as increasing shipping traffic of all kinds; with that will be a corresponding need for broadened naval

---

Council outside the convention, the only Arctic nation that is not part [of the convention], and one of the few nations still remaining outside one of the most widely subscribed international agreements in world history, we hinder our ability to lead. . . . We cannot stand outside the Convention and watch as other nations inside the convention accept the legal framework on issues of navigation, sovereignty, and resource rights that are critical to our nation. Having a seat at the table is extraordinarily important and it will diminish our maritime interests in the future if we do not subscribe to this." Communication to the committee from CAPT Timothy Gallaudet, USN, Deputy Director, Navy Task Force Climate Change, June 4, 2010.

[21]For example, the 2010 DOD Quadrennial Defense Review provides endorsement for U.S. ratification of UNCLOS in its discussion of climate and energy (see *Quadrennial Defense Review,* February 2010, p. 86 [page 108 of the PDF file], available at http://www.nationaljournal.com/congressdaily/issues/graphics/Defense-Review-2010.pdf). The committee realizes that the U.S. ratification of UNCLOS involves a number of nonmilitary issues. For additional reading, see Ronald O'Rourke, 2010, *Changes in the Arctic: Background and Issues for Congress*, Congressional Research Service, Washington, D.C., March 30, pp. 6-7; and National Intelligence Council, 1996, *Law of the Sea: The End Game, Intelligence Community Assessment,* March. Available at http://www.dni.gov/nic/special_endgame.html. Accessed November 23, 2009.

[22]White House (George W. Bush). 2005. *The National Strategy for Maritime Security,* Washington, D.C., September.

[23]See U.S. Department of Defense, *National Security and the UN Convention on the Law of the Sea,* Second Edition, 1996; available at http://www.dod.gov/pubs/foi/reading_room/876.pdf. Accessed July 28, 2010.

partnership and cooperation, and a framework for settling potential disputes and conflicts. By remaining outside the Convention, the United States makes it more difficult for U.S. naval forces to exercise maximum operating flexibility in the Arctic. Nonparticipation also complicates negotiations with partners for coordinated search and rescue operations in the region.

Beyond the Arctic, the committee anticipates increased HA/DR missions by U.S. naval forces as a result of projected increases in extreme climatic events. To support this potential increasing mission for humanitarian assistance to climate refugees and disaster-relief operations, allied partnerships will be essential. Hence the committee sees ratification of the Convention as an important national priority to leverage the enormous "soft power" of the treaty to share burdens and reduce the national security risks to the naval and joint forces and the nation. Becoming a party to the Convention then is clearly in the U.S. naval forces' best interests as the Arctic opens as a fifth ocean of interest.

FINDING 1.1: The committee has studied the implications of the failure of the United States to ratify the 1982 United Nations Convention on the Law of the Sea (UNCLOS) from the standpoint of potential impacts on national security in the context of a changing climate. As climate change affords increased access to the Arctic, it is envisioned that there will be new opportunities for natural resource exploration and recovery, as well as increased ship traffic of all kinds, and with that a need for broadened naval partnership and cooperation, and a framework for settling potential disputes and conflicts. By remaining outside the Convention, the United States makes it more difficult for U.S. naval forces to have maximum operating flexibility in the Arctic and complicates negotiations with maritime partners for coordinated search and rescue operations in the region.

RECOMMENDATION 1.1: The ability of U.S. naval forces to carry out their missions would be assisted if the United States were to ratify UNCLOS. Therefore, the committee recommends that the Chief of Naval Operations, the Commandant of the Marine Corps, and the Commandant of the Coast Guard continue to put forward the naval forces' view of the potential value and operational impact of UNCLOS ratification on U.S. naval operations, especially in the Arctic region.

NAVAL FORCES' POSITIONING ON CLIMATE CHANGE AND ENERGY

The leaders of the U.S. Navy, Coast Guard, and Marine Corps have recognized the potential impact of climate change on naval forces and have positioned their organizations to make adaptive changes.[24] For example, a joint Navy, Marine,

[24] A board of retired flag and general officers also recognized this impact and provided a broader perspective on the topic of national security and climate change. See Military Advisory Board, 2007, *National Security and the Threat of Climate Change*, CNA Corporation, Alexandria, Va.

and Coast Guard maritime strategy document identifies climate change as an area of concern and discusses areas for future potential naval attention.[25] In this regard, the CNO has recognized the linkage between energy use and climate change by establishing two key task forces: Navy Task Force Energy (charged with formulating a strategy and plans for reducing the Navy's reliance on fossil fuels—and thus reducing carbon dioxide emissions, operational energy demands, and, potentially, energy costs);[26] and Navy Task Force Climate Change (charged initially with developing a road map for Navy actions in the Arctic, and then with addressing longer-term Navy actions regarding global climate change policy, strategy, and plans).[27] This committee engaged with Navy Task Force Energy and Navy Task Force Climate Change and found that each is providing strong leadership on these issues across the Navy and DOD. Both task forces are well positioned in capability and credibility to continue their strong contributions.

Navy Task Force Climate Change issued its first report, U.S. Navy Arctic Roadmap, on November 10, 2009. The Arctic Roadmap is the first phase of a planned multistep approach for the U.S. Navy to address major climate change issues. The road map offers a chronological listing of Navy action items, objectives, and desired effects to address climate-change-related Arctic issues for the period FY 2010-FY 2014. Included in this Arctic road map are the following recommended FY 2011-2014 actions, actions that the Navy is reported to be acting upon or taking under consideration:[28]

- Initiate assessments of required Navy Arctic capabilities
- Develop recommendations to address Arctic requirements for program proposals in the Navy's Program Objective Memorandum for FY 2014 (POM-14)
- Continue biennial Navy participation in Arctic exercises, including ICEX-11, ICEX-13, Arctic Edge, and Arctic Care
- Formalize new cooperative relationships that increase Navy experience and competency in search and rescue (SAR), maritime domain awareness (MDA), and humanitarian assistance and disaster response in the Arctic, and defense support of civil authorities (DSCA) in Alaska.[29]

---

[25]See *A Cooperative Strategy for 21st Century Seapower,* Washington, D.C., 2007, p. 3. Available at http://www.navy.mil/maritime/MaritimeStrategy.pdf. Accessed July 28, 2010.

[26]CAPT James L. Brown, USN, Director, Navy Energy Coordination Office, Office of the Deputy Chief of Naval Operations for Fleet Readiness and Logistics, "Navy Task Force Energy, Perspectives and Related Climate Change Initiatives," presentation to the committee, September 17, 2009, Washington, D.C.

[27]See Vice Chief of Naval Operations (ADM W. Jonathan Greenert, USN), Memorandum 4000 Ser N09/9U103035, "Task Force Climate Change Charter," October 30, 2009.

[28]CAPT Timothy Gallaudet, USN, Deputy Director, Task Force Climate Change/Oceanographer of the Navy, "Task Force Climate Change Update and Gaps and Projected Future Needs," presentation to the committee, October 19, 2009, Washington, D.C.

[29]CAPT Timothy Gallaudet, USN, Deputy Director, Task Force Climate Change/Oceanographer of the Navy, "Task Force Climate Change Update and Gaps and Projected Future Needs," presentation to the committee, October 19, 2009, Washington, D.C.

As is evident in this report, the committee fully agrees with these initial actions and recommendations of TFCC. In developing future plans for Navy actions in and beyond the Arctic, TFCC has reported that it will draw upon the findings and recommendations from this report and other commissioned assessments.[30] Related, the U.S. Coast Guard (USCG) has commissioned the U.S. Coast Guard High Latitude Region Mission Analysis study to better define its specific requirements for protecting U.S. national security interests in the Arctic. The USCG study is anticipated to be completed in the third quarter of 2010.[31]

ADDITIONAL RELEVANT CLIMATE CHANGE ASSESSMENTS

Additional assessments on climate change and its impacts include the U.S. Global Change Research Program's 2009 report on *Global Climate Change Impacts in the United States.* Key findings of this government document were as follows:[32]

1. Global warming is unequivocal and is primarily human induced.
2. Climate changes are under way in the United States and are projected to grow.
3. Widespread climate-related impacts are occurring now and are expected to increase.
4. Climate change will stress water resources.
5. Crop and livestock production will be increasingly challenged.
6. Coastal areas are at increasing risk from sea-level rise and storm surge.
7. Threats to human health will increase.
8. Climate change will exacerbate many social and environmental stresses.
9. Thresholds will be crossed, leading to large changes in climate and ecosystems.
10. Future climate change and its impacts depend on choices made today.

Additionally, in response to a request from Congress, the National Research Council initiated the study America's Climate Choices, designed to inform and guide responses to climate change across the nation. Experts representing various levels of government, the private sector, nongovernmental organizations, and research and academic institutions populated panels on the following:

---

[30]The reader should note that Navy Task Force Climate Change was chartered to focus on issues for the U.S. Navy, whereas the terms of reference for this committee focus on issues impacting the U.S. Navy, Marine Corps, and Coast Guard.

[31]ADM Thad Allen, Commandant, U.S. Coast Guard, briefing to the committee, November 20, 2009, Washington, D.C.

[32]U.S. Global Change Research Program. 2009. *Global Climate Change Impacts in the United States,* Cambridge University Press, New York.

- Advancing the Science of Climate Change,
- Adapting to the Impacts of Climate Change,
- Limiting the Magnitude of Future Climate Change, and
- Informing Effective Decisions and Actions Related to Climate Change.

Reports from each of the first four America's Climate Choices panels are now available and should help inform future U.S. naval leadership decisions. For example, the *Advancing the Science of Climate Change* report recommends that a single federal entity or program be given the authority and resources to coordinate a national research effort integrated across many disciplines and aimed at improving both understanding and responses to climate change. The U.S. Global Change Research Program, established in 1990, could fulfill this role, but it would need to form partnerships with action-oriented programs and to address weaknesses in its current program. A comprehensive climate observing system, improved climate models and other analytical tools, investment in human capital, and better linkages between research and decision making are also essential to a complete understanding of climate change.[33] As discussed in Chapter 6 of this committee's report, the Navy is in position to both contribute to and benefit from such an effort.

PROPOSED NATIONAL AND GLOBAL FRAMEWORKS FOR CLIMATE SERVICES

In addition to needs expressed by the DOD and military services for better (decadal or longer) climate model projections, individuals and decision makers across widely diverse sectors—from agriculture to energy to transportation—increasingly are asking for information about climate change in order to make the best choices for their families, communities, and businesses. This translates to the provision of climate information on the regional level that investors, business leaders, natural resources managers, and policy makers need to help prepare for the adverse impacts of potential climate change on industries, communities, ecosystems, and nations. While global mean metrics of temperature, precipitation, and sea-level rise are convenient for tracking global climate change, many sectors of society require actionable information on considerably finer spatial and temporal scales—such as seasonal predictions with regard to Arctic sea ice and the seasonal prediction of hurricanes and other severe storms in specific areas of the world.

To meet the rising demand of these requests, the U.S. Commerce Secretary announced in February 2010 the intent to create a NOAA [National Oceanic and

---

[33]The initial four reports from the America's Climate Choices (ACC) studies are these: National Research Council, 2010, *Advancing the Science of Climate Change, Informing an Effective Response to Climate Change, Limiting the Magnitude of Future Climate Change,* and *Adapting to the Impacts of Climate Change,* The National Academies Press, Washington, D.C. Additional information on the ACC studies is available at http://americasclimatechoices.org. Accessed June 4, 2010.

Atmospheric Administration] Climate Service line office dedicated to bringing together the agency's strong climate science and service delivery capabilities.[34] One of the challenges to be faced by an effective Climate Service is the sheer interdisciplinary breadth of providing climate services across such sectors as agriculture, parks and recreation, terrestrial ecosystems, insurance and investment, energy, state/local/municipal governments, water, human health, commerce and manufacturing, transportation, and coastal and marine sectors.

The impact of climate on many of these sectors has direct implications for national security. Internationally, a question of paramount importance confronting nations is how to adapt to the prospect of climate variability and change in the next half century. In response, the World Climate Conference-3 convened in Geneva, Switzerland, in 2009 to establish a Global Framework for Climate Services that addresses the needs of decision makers worldwide for accurate and timely climate information and predictions. Delegations from 163 nations met in Geneva to ensure that current and future generations have access to the climate predictions and information necessary for various socioeconomic sectors to cope with climate variability and change.[35]

RISK MANAGEMENT IN THE FACE OF FUTURE CLIMATE CHANGE PROJECTIONS

This committee believes that there is strong scientific evidence to recommend that U.S. naval leadership should continue their ongoing attention to the national security implications of climate change, specifically its potential impact on future naval operations and capabilities. However, current scientific understanding and predictive capability for climate change lack the specificity that the Navy needs for planning purposes. Deficits in knowledge include, for example, the rate of future sea-level rise, the timing for the opening of Arctic waters, and reliable predictions of regional climate (given the current inability to project specific regional impacts). Considering that it is unlikely that the precision of climate change projections will dramatically improve in the next few years, the committee believes that the Navy should adopt a risk management approach for addressing these issues. Such an approach should include a range of contingency plans for the potential onset of climate-driven severe-weather disasters.

[34]See *Commerce Department Proposes Establishment of NOAA Climate Service*, NOAA news release, February 8, 2010; available at http://www.noaanews.noaa.gov/stories2010/20100208_climate.html. Accessed June 4, 2010.

[35]Participants at the World Climate Conference-3 approved, by acclamation, a conference declaration deciding to establish a Global Framework for Climate Services (GFCS) to strengthen production, availability, delivery, and application of science-based climate prediction services, and they outlined a path forward for establishing the GFCS. "Summary of the World Climate Conference," *World Climate Conference Bulletin,* Vol. 165, No. 1, September 2009. Available at http://www.iisd.ca/ymb/climate/wcc3/html/ymbvol165num1e.html. Accessed June 4, 2010.

Some recent reports have also noted the possibility of "climate surprises," that is, unexpected rapid changes outside of current climate model projections.[36] Such surprises would likely be associated with fast processes and interactions/feedbacks among different Earth system components, including physical, chemical, and biological aspects. Naval operations could be particularly affected by these "surprises" if there is abrupt acceleration of sea-level rise, rapid sea-ice loss, rapid thawing of the permafrost releasing additional CO_2 or methane into the environment, or an increase in strong tropical cyclones. However, means for reliably predicting such abrupt changes are not currently available.

The committee discussed, but did not explore in depth, risk management for climate change planning. The committee is aware, however, of a growing body of knowledge in this area. These risk analysis approaches examine means for conceptualizing and managing risk for "fat-tailed distribution" events, which may exhibit abrupt change or have high impact and are not mathematically well behaved.[37] While the insurance industry is the target and sponsor for many of these examinations, the principles and issues explored are applicable to naval climate change risk analyses.[38]

Many uncertainties remain about the course of climate change, and these will probably continue in the near term. It is noteworthy that the U.S. Navy's assets and entities, including the Office of Naval Research, the Naval Research Laboratory, and the Naval Meteorology and Oceanography Command, are recognized by climate scientists as critical partners in advancing the understanding of climate science and related policy implications.[39] The committee strongly supports the continuation of dedicated efforts by the Navy to remain engaged with and assist

[36]Thomas R. Karl, Gerald A. Meehl, Susan J. Hassol, Anne M. Waple, and William L. Murray (Eds.). 2008. *Weather and Climate Extremes in a Changing Climate,* Department of Commerce, NOAA's National Climatic Data Center, Washington, D.C.

[37]In statistics, the term "fat-tail distribution" is used to describe the probability of high-consequence events that fall on the tail end of a statistical distribution and cannot be accurately described by the normal distribution bell-shaped curve. In these cases, the probability of an extreme event, though unlikely, is higher than it would have been under the normal hypothesis and assignment of risk.

[38]For example, see Carolyn Kousky and Roger M. Cooke, 2009, "Climate Change and Risk Management: Challenges for Insurance, Adaptation and Loss Estimation," discussion paper, Resources for the Future, Washington, D.C., February. Available at http://www.rff.org/documents/RFF-DP-09-03.pdf. Accessed July 28, 2010.

[39]U.S. Navy and Coast Guard assets have been highly important in providing critical scientific data associated with both ice mass and ocean changes over extended periods. Also, the Measurements of Earth Data for Environmental Analysis (MEDEA) Program, a project of the 1990s, has been highly valuable in providing sea-ice data from military and intelligence assets that would otherwise be unavailable in the civilian sector. See National Research Council, 2009, *Scientific Value of Arctic Sea Ice Data,* The National Academies Press, Washington, D.C. In another example, Scientific Ice Expeditions (SCICEX) was a 5-year program in which the Navy made available a Sturgeon-class nuclear-powered attack submarine for unclassified science expeditions to the Arctic Ocean to gather ice-thickness measurements. Additional information on SCICEX is available at http://www.nsidc.org/noaa/scicex/. Accessed May 10, 2010.

in leading advancements in climate science and understanding its impacts within the broader context of the DOD's responsibility to assess the effects of climate change on all DOD missions, capabilities, and facilities. The Navy brings significant historical experience and unique assets to this arena, such as its specialized oceanographic fleet and its submarine under-ice data collection capabilities. The committee views these naval assets and related advances in fundamental knowledge as supportive of the national security interests of the United States. These advances will also aid in reducing the uncertainties in future climate change and response projections and the necessary national response.

FINDING 1.2: Many climate change issues with potentially high impact for U.S. naval forces are known through direct scientific evidence. However, many climate change uncertainties remain.

RECOMMENDATION 1.2: Naval leadership should continue leading the understanding of the impact of climate change for the DOD and contribute to better technical understanding. Naval leadership should adopt a risk analysis approach for dealing with the climate change uncertainties by developing response scenarios for accepted projections and for excursions away from the projections.

Throughout this report, the committee recommends actions that would address these more extreme excursions as a hedge against the possible and to help address a risk management approach.

ORGANIZATION OF THE REPORT

This report is organized to address the key naval issues requested in the study's terms of reference and presents findings and recommendations in the six areas for naval leadership action outlined in the report Summary. Following the Chapter 1 introductory comments, Chapter 2 addresses national security and climate-change-related operational issues for U.S. naval forces.[40] Where appropriate, Arctic operational issues for naval forces are articulated separately from global climate-change-related naval operational issues. Chapter 3 addresses physical infrastructure issues for U.S. naval forces, due primarily to projected climate-induced sea-level rise and storm surge and their potential impact on naval coastal installations around the globe. Allied forces issues are discussed and explored in Chapter 4, not only providing findings and recommendations for climate change issues associated with traditional U.S. allies but also addressing the need for broader naval military and international partnerships in planning for projected climate-induced HA/DR events. Chapter 5 discusses climate-change-

[40]Throughout the report, unless otherwise designated, the term "naval forces" includes the U.S. Navy, the U.S. Marine Corps, and the U.S. Coast Guard.

related technical issues for U.S. naval forces, including addressing antisubmarine warfare in a changing world ocean. Where appropriate, Chapter 5 also explores Arctic technical challenges separately from global climate-change-related technical challenges. Chapter 6, the concluding chapter of the report, provides an examination of future research and development needs and associated findings and recommendations for U.S. naval forces—concentrating on those areas in which the committee believes the naval forces have particular interests that might not likely be met in the near term by other groups pursuing climate science research.

2

Naval Capabilities and Potential Climate-Change-Related Operational Issues

INTRODUCTION

This committee has found strong scientific evidence to support naval leadership's initiatives to study and act on the implications of climate change and its effects on naval missions, operations, and capabilities. Numerous peer-reviewed assessments indicate increasing global stresses due to the effects of climate change alone and in combination with other environmental stressors, such as global population growth.[1,2] These reports and scientific models suggest more severe or frequent droughts, floods, storms, and other events with negative consequences for

[1]In many regions of the world, the impact of climate change is likely to further exacerbate the preexisting stress on water supplies and the mounting pressures of population growth. For example, Columbia University's Center for International Earth Science Information Network (CIESIN) has compiled information from Intergovernmental Panel on Climate Change (IPCC) assessments, the 2005 World Bank report *Natural Disaster Hotspots: A Global Risk Analysis,* and CIESIN's gridded world population data sets to present a projected geographic distribution of vulnerability in 2050. In presentations to the committee, CIESIN representatives reported that global population nearly doubled from 1968 through 2008, and that by 2048 it could grow another 40 percent, to more than 9 billion people, adding even greater stresses to water and food supplies. CIESIN also reports that population increases are fastest in areas most vulnerable to intense storms and flooding (e.g., coastal areas, islands, and river basins). The CIESIN analysis combines its population data sets with IPCC-projected climate-change-related vulnerabilities, economic data, and past disaster-related losses to identify areas at relative high risk from one or more hazards. See Robert S. Chen, Center for International Earth Science Information Network, Columbia University, "Human Dimensions of Climate Change," and Marc Levy, Center for International Earth Science Information Network, Columbia University, "Climate Change and U.S. National Security," presentations to the committee, November 19, 2009, Washington, D.C.

[2]For example, see Thomas R. Karl, Jerry M. Melillo, and Thomas C. Peterson, 2009, *Global Climate Change Impacts in the United States,* Cambridge University Press, New York.

food and water supplies, possibly leading to even greater stress on the expanded human population.[3]

This chapter begins with an examination of climate change impacts on naval forces' missions and operations—including increased humanitarian assistance/disaster relief (HA/DR) and the resulting implications for such units as U.S. Navy hospital ships, Naval Mobile Construction Battalions (NMCBs), and Marine Expeditionary Units (MEUs). The report then focuses on climate-change-related operational impact and challenges in the Arctic, highlighting Arctic command issues and an examination of U.S. icebreaker needs. This chapter concludes with a discussion of the implications of a changing climate on health and disease and the impact that this may have for future naval missions.

Viewed from a national security standpoint, the above changes would likely amplify stresses on weaker nations and generate geopolitical instability in already vulnerable regions.[4] A range of naval mission impacts may result from such conditions, including the sorts of antipiracy and counterterrorism missions now being conducted off Somalia. However, the clearest implications are for a potential increase in the frequency of HA/DR missions. These additional HA/DR demands have the potential to strain military transportation resources and supporting force structures.

The U.S. Navy, as a forward-deployed force, is in position to reach disaster relief sites faster than other agencies and will almost assuredly experience increased demand for assistance if disasters increase due to climate change.[5] The

[3]See Intergovernmental Panel on Climate Change, 2007, "Climate Change 2007: The Physical Science Basis," Working Group I contribution to the *Fourth Assessment Report of the Intergovernmental Panel on Climate Change* (Susan Solomon, Dahe Qin, Martin Manning, Zhenlin Chen, Melinda Marquis, Kristen B. Averyt, Melinda M.B. Tignor, and Henry LeRoy Miller [eds.]), Cambridge University Press, Cambridge, United Kingdom, and New York. See also Catherine P. McMullen and Jason Jabbour, 2009, *Climate Change Science Compendium,* United Nations Environment Programme, EarthPrint, Nairobi, Kenya.

[4]See Statement of the Record of Dr. Thomas Fingar, Deputy Director of National Intelligence for Analysis and Chairman of the National Intelligence Council, before the Permanent Select Committee on Energy Independence and Global Warming, House of Representatives, "National Intelligence Assessment on the National Security Implications of Global Climate Change to 2030," June 25, 2008. Available at http://www.dni.gov/testimonies/20080625_testimony.pdf. Accessed November 24, 2009. See also Military Advisory Board, 2007, *National Security and the Threat of Climate Change*. CNA Corporation, Alexandria, Va.

[5]*Naval Operations Concept 2010* (NOC 10)—a joint maritime strategy document for the U.S. Navy, Marine Corps, and Coast Guard—calls out "humanitarian assistance and disaster response" as one of six capabilities that constitute the core of U.S. maritime power and that "reflect an increased emphasis on those activities that prevent war and build partnerships." See Department of the Navy and U.S. Coast Guard (ADM Gary Roughead, USN; Gen James T. Conway, USMC; and ADM Thad W. Allen, USCG), 2010, *Naval Operations Concepts 2010, Implementing the Maritime Strategy,* June. Available at http://www.navy.mil/maritime/noc/NOC2010.pdf. Accessed June 4, 2010. However, it is not the sole responsibility of the U.S. military to respond to national and international humanitarian and disaster-relief emergencies; many U.S. and international governmental and private agencies may be engaged in any given relief operation.

demand for Naval Construction Force capability in support of HA/DR operations is likely to increase in proportion to the operational tempo of U.S.-sponsored international HA/DR operations.[6] Likewise, the U.S. Marine Corps, with its forward-deployed MEUs, should expect to be called upon to assist with extreme-weather-related HA/DR. However, the pace and extent of this increase are as yet unknown.

The committee sees three fundamental challenges facing U.S. naval forces regarding climate change impacts on missions, capabilities, and operations:

- The need to develop capabilities, including logistics and training, to support new missions that climate change may bring;
- The need to respond to an increase in the demand for certain types of existing missions; and
- The need to maintain current warfighting capabilities as the operating environment changes.

Regarding new or expanding missions, the committee considers the need to operate in the Arctic and the expected increase in demand for HA/DR missions and operations related to mass migrations to be most likely. Regarding the maintenance of current capabilities in a changing operational environment, the ability of the Navy to project power under harsher climate conditions and the robustness of its antisubmarine warfare (ASW) capability as the acoustic environment changes are among the major issues. Each of these challenges is discussed below. ASW and other technical operational issues are discussed more fully in Chapter 5 of this report.

NAVAL FORCES' RESPONSES TO FUTURE POTENTIAL CLIMATE-INDUCED EVENTS

Humanitarian Assistance/Disaster Relief

All U.S. military services and many other federal agencies could be involved in supporting HA/DR missions brought on by climate change, depending upon the nature of the crisis, its location, and the severity of the event. Forward-deployed naval forces (Navy, Marines, and Coast Guard) are likely to be in the best position to respond rapidly to developing HA/DR crises and are therefore very likely to be called upon by the President. It is also probable that naval forces of coalition partners would be involved as part of the effort to bring relief to the affected area. Examples of international HA/DR efforts are the 2010 earthquake in Haiti, Tropi-

[6] For a review of U.S. Navy Construction Battalion operations, see U.S. Navy Seabees First Naval Construction Division, Strategic Plan 2008-2011, Norfolk, Va.

cal Storm Ketsana striking the Philippines in 2009, and the tsunami in Indonesia in 2004.

Navy forces afloat, Amphibious Ready Groups (ARGs) with embarked MEUs, and Maritime Prepositioning Force (MPF) squadrons can all bring a unique level of rapid response capability in support of Combatant Commander (COCOM) requirements for HA/DR missions that are over and above traditional warfighting capability. The most recent example of this would be the deployment of the aircraft carrier USS *Carl Vinson,* which operated as a sea base for helicopters that were moving personnel and supplies into the disaster area in Haiti, in addition to two ARGs with embarked MEUs. In the future, other Navy surface elements—such as littoral combat ships (LCSs), joint high-speed vessels, and Navy support ships—might also play a role depending on the nature of the mission. At this point in time, it is difficult to project what these relief forces may look like in the decades ahead; however, the current organizational structure as well as the platforms may be essentially the same.

Hospital Ships

The U.S. Navy's hospital ships are one of the best examples of Navy "soft power" as the United States faces an uncertain future in the 21st century. Navy hospital ships have played a significant role in HA/DR missions in the Pacific as well as the Caribbean, and they will continue to do so as long as they are in active service. Their unmatched medical capacity and capability utilized in support of COCOM requirements could be increasingly in demand should the effects of climate change create circumstances in which massive medical assistance is needed. However, manning these platforms with the requisite medical expertise may require an innovative program whereby staffing comes from the United States, the host nation, or nongovernmental organizations (NGOs) on a contingency basis. Most importantly, while to the knowledge of this committee no retirement of these hospital ships has been officially discussed, the capability that these ships bring to the COCOM must be preserved in some form when they are eventually decommissioned. Without the hospital ships, the requirement to provide medical support would reside with medical capabilities located in large-deck amphibious ships (LHA/LHD classes), aircraft carriers, and possibly the LPD-17-class amphibious ships. It could require several of these ships to be in the area of HA/DR operations to match the capacity of a single hospital ship. That may not always be feasible due to COCOM requirements elsewhere in the world.

Another related issue involves the use of host-nation medical teams and NGOs on U.S. ships. While this integration has been successful on U.S. hospital ships, it will not work so easily or smoothly on U.S. combatant or amphibious ships, due to the nature of these ships' primary missions.

Maintaining the capabilities provided by the hospital ships goes beyond just ensuring an equivalent bed capacity. The Haiti earthquake disaster demonstrated the increasing need for timely and sophisticated surgical/trauma care. Many of the land-based hospitals—including those brought in by other governments and U.S. NGO groups—could not handle severely injured patients. Fortunately, within a few days of the arrival of USNS *Comfort,* protocols were developed for the transfer of the most seriously injured.[7] It will be important in future planning for HA/DR to focus not only on the number of beds but also on the number of operating rooms and provision for sophisticated trauma care. Equally important will be the speed with which the services can be on station.

One potential solution to the requirement for hospital ships for HA/DR could be the use of contracted afloat services, potentially provided by large, private shipping companies. The committee has reviewed reports on such an option studied as part of the Chief of Naval Operations (CNO)-directed study of sea basing options, or the afloat forward staging base concept. This committee believes that such a potential third-party arrangement deserves consideration as the Department of Defense (DOD) and naval forces study their options for HA/DR support and capacity. Preliminary projections of a hypothesized converted commercial ship's capabilities are as follows: command, control, communications, and computers—capabilities robust enough to support the on-scene commander and staff; berthing/hotel facilities for 3,000 responders; medical facilities that range from operating rooms to a 1,000-bed hospital ward; potable water making/bottling; ice making; transport and discharge for wheeled responder vehicles; transport for amphibious response craft; aviation facilities (including refueling) for all rotary wing aircraft (up to 14 landing spots); and an enormous amount of cargo space for food, medical supplies, and the like.[8] Such a potential third-party commercial ship-leasing arrangement might also be based on shared-cost bilateral or multilateral HA/DR agreements and might possibly bring cost savings benefits. Any such arrangements must meet all requirements in a short time frame. The length of time between injury and care is a critical variable in medical outcome.

FINDING 2.1: The unique capability provided by the U.S. Navy hospital ships will become even more important in supporting potential humanitarian assistance/disaster relief (HA/DR)-related missions that will likely occur as a result of crises created by climate change. The Navy needs to maintain this capability beyond the life of its current two-ship hospital fleet.

RECOMMENDATION 2.1: The Program Executive Office for Ships (PEO-Ships), the Naval Sea Systems Command (NAVSEA), and the Military Sealift

[7]Paul S. Auerbach, Robert L. Norris, Anil S. Menon, Ian P. Brown, Solomon Kuah, Jennifer Schwieger, Jeffrey Kinyon, Trina N. Helderman, and Lynn Lawry. 2010. "Civil-Military Collaboration in the Initial Medical Response to the Earthquake in Haiti," *New England Journal of Medicine,* Vol. 362:e32, No. 10, March 11.

[8]Private communication, Robert Bowers, Senior Director, Maritime Technical Services, Maersk Line, Limited, with ADM Frank Bowman, USN (Ret.), committee co-chair, July 24, 2009.

Command (MSC) should analyze alternatives to retain the medical capability of the current hospital ships into the future. The analysis should address construction of new military or commercial platforms like the Mobile Landing Platform (MLP) that will join the Maritime Prepositioning Force (MPF); modification to current surface platforms or amphibious "big-decks"; or construction of next-generation Navy fleet hospitals to meet the requirements. In this context, PEO-Ships, NAVSEA, and MSC should also explore the feasibility of leasing commercial ships and crews to meet the requirements, but in doing so must ensure that the provisions for operating rooms, sophisticated trauma care, and guaranteed availability on very short notice are included.

Marine Expeditionary Units

There are usually at least three Amphibious Ready Groups with their Marine Expeditionary Units forward-deployed on presence missions at all times in support of COCOM requirements. Each MEU is trained in a variety of noncombat missions to include HA/DR, noncombatant evacuation orders, and security force operations.

Each MEU maintains 15 days of sustainability in a self-contained sea base. The large-deck amphibious ship in each ARG has a medical capability as well as a platform for sustained helicopter and MV-22 military transport aircraft operations that can operate independently or in conjunction with other naval elements, including a hospital ship. It can respond rapidly to a crisis area and also function as the command and control headquarters until a shore-based force arrives on the scene. If the nature of the HA/DR crises requires the commitment of additional naval forces, decisions can be made to flow two or possibly three of the deployed ARGs/MEUs to the area in question. This force could merge into a larger Marine Expeditionary Brigade (MEB) that would conduct operations from an expanded sea base with some 45 days of sustainability plus embarked engineer assets for light engineering tasks. At some point, and as part of the committee's recommendation for risk analysis, naval planners may need to develop contingency plans in preparation for such deployments as events created by climate change become clearer in specific regions of the world. Logistics requirements for sustaining the force as well as supply and medical assistance for the affected population in the area of operations could be very different from what is routine for conventional operations today.

The Maritime Prepositioning Force

The MPF, which is forward-deployed with elements of the fleet, can provide a significant logistics capability that can move rapidly to an area in the event of an HA/DR crisis. This force, which is composed of three squadrons of five ships each, is generally located in the Pacific, the Indian Ocean, and the Mediterranean.

Each squadron contains a 30-day supply for an MEB; the support for one Naval Mobile Construction Battalion, or Seabees; materials for a 3,700-foot runway and control equipment for an expeditionary airfield (EAF); and a Navy fleet hospital (NFH) with a 273-bed capacity that can be deployed ashore. Each MPF squadron contains 374,000 gallons of bulk water and can produce 122,000 gallons of water per day that can be moved ashore. One of these MPF squadrons could become the logistics hub for an expanded sea base that can support Marine and NMCB operations ashore as well as provide assistance for NGOs, as was done in Haiti. If a major crisis like mass migration, conflicts over water, or a natural disaster occurs, decisions could be made to put a second MPF squadron into an area for extended operations from both the sea base and ashore.

Each MPF squadron has a Naval Support Element (NSE) that has the mission of facilitating the off-loading of MPF shipping in stream or pierside. Within the NSE are several units, including an Amphibious Construction Battalion that has the capacity to build its own camp as well as move equipment ashore. This particular organization of Seabees could be utilized to assist in HA/DR operations if the requirement for assistance expands ashore and inland.

Naval Mobile Construction Battalions

The Navy's NMCB units are capable of a wide range of heavy construction and should be considered a national asset that would be available in response to HA/DR missions or crises brought on by the effects of climate change.[9] Each NMCB unit is self-contained, with its own support structure, and it can provide an expeditionary brigade with a wide range of construction capabilities to include site preparation, roads, airfields, and buildings if necessary. One Seabee unit will support a Navy fleet hospital ashore as well as construct the expeditionary airfield if there is a requirement to build one in an area of operations. The NMCB has the capability for drilling wells for water that might be critical if a crisis involves mass migration or famine as a result of severe drought. The role of the Seabees could increase considerably if there is a requirement to move operations inland in order to manage the crises. Additional Naval Mobile Construction Battalion units can also be flown into the area of operations to expand current capabilities if the need arises.

The Coast Guard

The U.S. Coast Guard's (USCG's) core roles are to protect the public, the environment, and the U.S. economic and security interests in any maritime region in which those interests may be at risk, including international waters and

---

[9]For a review of U.S. Navy Construction Battalion operations, see U.S. Navy Seabees First Naval Construction Division, Strategic Plan 2008-2011, Norfolk, Va.

America's coasts, ports, and inland waterways.[10] In doing so, the Coast Guard has organized its responsibilities into five fundamental areas: (1) maritime safety, (2) national defense, (3) maritime security, (4) maritime mobility, and (5) protection of natural resources, and a unique mission in ice operations in which icebreakers play a key role. The Coast Guard will play a role in HA/DR operations, especially in the Caribbean and Eastern Pacific regions. It will often be the primary responder to a natural disaster or mass migration in North America, as recently seen in Hurricane Katrina (2005) and the Haitian earthquake (2010). The Coast Guard has the initial responsibility to manage mass migrations by sea, primarily in the Caribbean, until the number of migrants requires the Navy to provide support. The Coast Guard will also provide expertise for port operations, aids to navigation, maritime search and rescue, and vessel traffic control.

Mass Migrations

The CNA study of 2007[11] and the 2010 *Quadrennial Defense Review* (QDR)[12] noted that climate change is likely to be an accelerant of instability. A similar document from the United Kingdom Ministry of Defence[13] noted that climate change is "likely to be most severe where it coincides with other stresses such as poverty, demographic growth and resource shortages." Increasing mass migrations can occur as a result of resource shortfall, conflict, or sea-level rise, all of which are likely to increase with climate change. Recent estimates are that in 2050 the world's population will be 9 billion people, and that 200 million people could be newly mobilized as climate migrants due to climate change effects.[14] These mass migration increases will not only affect the humanitarian assistance requirements of the naval forces, but could also result in instability as well as unrest and regional conflict. One possible outflow of these events is the evacuation of U.S. citizens in the impacted region and rendering of assistance in quelling unrest.

Regarding migration driven by sea-level rise, southern Asia (particularly Bangladesh) and Africa are most often mentioned; however, one of the great questions facing the international community is the potential disaster of small island nations that may disappear completely as sea level rises. This is an extreme, but nonetheless potential, scenario that may require naval evacuations. Small island

[10]See U.S. Coast Guard Missions, available on the U.S. Coast Guard/U.S. Department of Homeland Security website at http://www.uscg.mil/top/missions/. Accessed July 28, 2010.

[11]Military Advisory Board. 2007. *National Security and the Threat of Climate Change*, CNA Corporation, Alexandria, Va.

[12]Secretary of Defense (Robert M. Gates). 2010. *Quadrennial Defense Review,* Department of Defense, Washington, D.C., February, p. xv.

[13]"Adaptability and Partnership: Issues for the Strategic Defence Review," presented to Parliament by the Secretary of State for Defence, February 2010.

[14]Michael Werz and Karl Manlove. 2009. "Climate Change on the Move: Climate Migration Will Affect the World's Security," Center for American Progress, Washington, D.C., December 8.

nations are already losing freshwater resources as a result of salt water intrusion from a rising ocean.

The primary responsibility for dealing with (maritime) mass migrations to the United States now rests with the Coast Guard and the State Department. However, when the level of migrants reaches 1,000 per day, the Navy is called upon to assist the Coast Guard.

Plans, Training, and Provisioning of Forces

As the committee reviewed the impacts of climate change on the operations of naval forces, it became apparent that changes in mission, increased operations in existing missions, and operating in the new environment that might be expected in 20 years' time should affect how the Navy plans, trains, and equips its forces. The committee suggests that planning scenarios be revised to include climate change effects, and war gaming be conducted to test the functionality of the plans in light of the new challenges to operations. The lessons learned from these war games can then be used to review the adequacy of current force structure and training to meet the future challenges presented by climate change. This should result in a gap analysis and changes to the required skills and capabilities that drive force planning. A similar gap analysis will provide insight on the current provisioning and equipping of the various forces, particularly the contingency forces that will most likely be utilized in responding to HA/DR missions. Finally, the adequacy of logistical support for new and increased operations should be reviewed and the appropriate resources devoted to modifying policy for and funding of logistical support.

FINDING 2.2: Global climate change projections from the Intergovernmental Panel on Climate Change *Fourth Assessment Report* (AR4) suggest damaging impacts in developing and developed nations that may be destabilizing in many parts of the world. These projections would affect U.S. national security and stress naval resources. In particular, naval forces will likely be required to carry out more frequent humanitarian assistance/disaster relief (HA/DR)-related missions. At the same time, U.S. naval forces would be expected to execute their ongoing national security military missions and to position themselves for supporting missions in destabilized regions around the globe. It is also expected that the demand for U.S. Naval Construction Force and Marine Expeditionary Unit capabilities will increase in proportion to the operational tempo of U.S.-sponsored international HA/DR missions.

RECOMMENDATION 2.2: In the near term, the Chief of Naval Operations (CNO) should not specifically fund new force-structure capabilities to deal with the effects of projected climate change; however, the CNO should begin to hedge against climate change impacts through planning for modifications of the existing force structure as climate change requirements become clearer. The U.S. naval forces (the U.S. Navy, Marine Corps, and Coast Guard) should begin to consider potential specific force-structure capabilities and training standards for conducting missions arising from, or affected by, climate change, particularly HA/DR-related missions.

Naval Capabilities and Potential Climate-Change-Related Operational Issues in the Arctic

Changes in Arctic Ice Cover and Its Implications

Recent climate change may have the most immediate and obvious implications for maritime operations in the Arctic region.[15] The Arctic is experiencing significant reductions in sea-ice cover in the Arctic Ocean and the disappearance of older, thicker, multiyear ice.[16] The loss of sea-ice area in summer months is about three times faster than in winter. As a result, the vast Arctic is rapidly acquiring the types of maritime activities that normally occur elsewhere in the world's ice-free oceans.

Projected sea-ice retreat will offer a longer season of maritime availability. At the same time, community resupply demands are expected to rise with increasing development, migration, and population growth.[17] The U.S. Geological Survey notes that significant natural resources (oil, natural gas, and nonfuel minerals) may become increasingly available for exploitation as ice melts and climate tempers.[18] Tourism is expanding, especially around Greenland and Svalbard, but also in recent years in the Northwest Passage and around Arctic Alaska. There is evidence that commercially valuable fish stocks are moving north, and although U.S. waters north of the Bering Strait have been closed to fishing for the immediate future, the

---

[15]As discussed in Chapter 1, in this report the Arctic region is defined as the land and sea area north of the Arctic Circle.

[16]On September 12, 2009, sea-ice extent reached a 2009 minimum of 5.1 million km^2. The summer minimum is the third-lowest recorded since 1979. While the 2009 minimum was an increase over that of the two previous years, it was still 1.6 million km^2 below the 1979-2000 average minimum. The March 2009 ice extent was 15.2 million km^2, the same as in 2008 and only 4 percent less than the 1979-2000 average of 15.8 million km^2. March is historically the month of maximum sea-ice extent. See *Arctic Report Card: Update for 2009;* available at http://www.arctic.noaa.gov/reportcard/ArcticReportCard_full_report.pdf. Accessed November 24, 2009. See also Julienne A. Maslanik, C. Fowler, J. Stroeve, S. Drobot, J. Zwally, D. Yi, and William J. Emery, 2007, "A Younger, Thinner Arctic Ice Cover: Increased Potential for Rapid, Extensive Sea-Ice Loss," *Geophysical Research Letters,* Vol. 34, L24501.

[17]Arctic Council. 2009. *Arctic Marine Shipping Assessment 2009 Report;* available at http://www.nrf.is/index.php/news/15-2009/60-arctic-marine-shipping-assessment-report-2009. Accessed November 24, 2009.

[18]See July 23, 2008, U.S. Geological Survey press release, "90 Billion Barrels of Oil and 1670 Trillion Cubic Feet of Natural Gas Assessed in the Arctic"; available at http://www.usgs.gov/newsroom/article.asp?ID=1980&from=rss_home. Accessed November 23, 2009.

fishing area may expand.[19] A map of the Arctic region is shown in Figure 2.1; a profile of recent monthly Arctic sea-ice extent is provided in Figure 2.2.

Climate model projections are especially uncertain at regional scales and in regions with very rapid projected change such as the Arctic. The sea ice in most climate models retreats more slowly than what has been observed during the satellite era (since 1979).[20] Of the two models that are consistent with the satellite observations, one shown in Figure 2.3 projects an open Northern Sea Route in August and September and an approximate 50 percent ice concentration in the months of July and October by 2030. This same model has a significant rise in the year-to-year variability in sea-ice cover as the sea ice retreats in the 21st century.[21] This model suggests that large anomalies in the sea-ice cover as observed in 2007 may be increasingly common as the sea ice continues to retreat. Hence, these models support the likelihood of an Arctic maritime area increasingly accessible to surface shipping.

This committee believes that U.S. naval leadership should expect the decline in Arctic summer sea ice to continue at the current rate (10 percent per decade) or more in the next few decades. This would allow "ice-free" access over large stretches of the Arctic in late summer by 2030 that are sufficient for reliable cross-Arctic transit.[22] (See also Figure 2.3, and Figure 6.2 in Chapter 6 of this report.) In the near term, ice-laden Arctic waters will continue to have an ice cover of variable thickness and duration and will continue to pose navigational hazards for non-ice-hardened vessels.

Geopolitical and Military Issues

Essentially ignored since the end of the Cold War, the geopolitical situation in the Arctic has become complex and nuanced. Of the five nations that border the Arctic Ocean—Canada, Denmark (by virtue of its responsibilities for Greenland),

---

[19] Arctic Council. 2009. *Arctic Marine Shipping Assessment 2009 Report;* available at http://www.nrf.is/index.php/news/15-2009/60-arctic-marine-shipping-assessment-report-2009. Accessed November 24, 2009.

[20] Julienne Stroeve, Marika M. Holland, Walt Meier, Ted Scambos, and Mark Serreze. 2007. "Arctic Sea Ice Decline: Faster Than Forecast," *Geophysical Research Letters,* Vol. 34.

[21] Marika M. Holland, Cecilia M. Bitz, L.-Bruno Tremblay, and David A. Bailey. 2008. "The Role of Natural Versus Forced Change in Future Rapid Summer Arctic Ice Loss," *Geophysical Monograph Series 180,* American Geophysical Union.

[22] Throughout this report, the term "ice-free" is used to mean that multiyear ice has nearly (or completely) disappeared; however, to date, in what is termed "ice free" conditions, sufficient ice is present to remain a hazard to ordinary ships and routine marine operations. The Navy Task Force Climate Change also uses a projection of ice-free summer months in the Arctic by the year 2030 based on work conducted for the Department of Defense by the Oak Ridge National Laboratory using outputs from the Community Climate System Model version 3 (CCSM3). See Karsten Steinhaeuser, Esther Parish, Alex Sorokine, and Auroop R. Ganguly, 2009, "Projected State of Arctic Sea Ice and Permafrost by 2030," Oak Ridge National Laboratory, Oak Ridge, Tenn.

FIGURE 2.1 Map of the Arctic region. The Arctic region, in this report, is defined as the area north of the Arctic Circle (highlighted on this map in red).

Norway, Russia, and the United States—all are NATO members except Russia. The Arctic Council, a governmental forum of these five nations plus Iceland, Sweden, and Finland, offers a diplomatic vehicle for addressing contemporary Arctic issues. However, maritime boundary disputes abound. Canada and the United States, and Canada and Denmark, have unresolved territorial sea and exclusive economic zone (EEZ) disputes in the Arctic. Norway and Russia disagree over

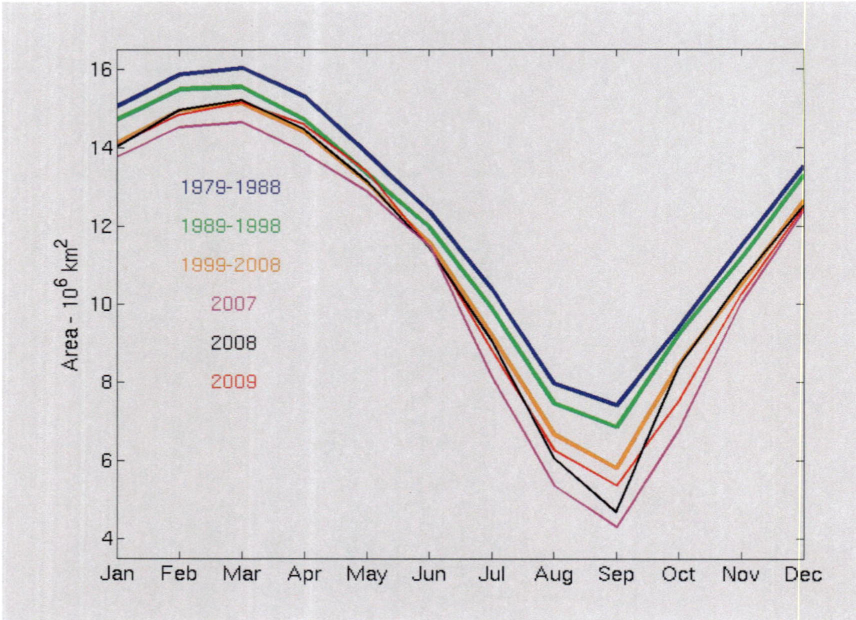

FIGURE 2.2 Sea-ice-extent data by month for 2007 through 2009 with decade averages to illustrate the trends by month. The year 2007 was the record low in nearly every month. Sea-ice extent during 2008 and 2009 recovered toward the average of the 1999-2008 decade. Data are available at http://nsidc.org/data/docs/noaa/g02135_seaice_index/. SOURCE: F. Fetterer, K. Knowles, W. Meier, and M. Savoie. 2002, updated 2009. Sea Ice Index. Boulder, Colorado, USA: National Snow and Ice Data Center. Digital media.

offshore areas around Svalbard. The status of the Northwest Passage through the Canadian archipelago—internal Canadian waters or an international strait—has been a Canadian concern since at least 1985. The issue is not resolved. Currently, icebreaker transits are allowed through nation-to-nation bilateral agreements.[23] The most notable issues involve existing and potential claims of the extended outer continental shelf under provisions of the United Nations Convention on the Law of the Sea (UNCLOS). Russia's dramatic planting of a titanium flag on the Arctic Ocean sea bottom at the North Pole in 2007 prompted a U.S. policy review resulting in National Security Presidential Directive (NSPD)-66/Homeland Security Presidential Directive (HSPD)-25, and raised the possibility that Arctic issues will require national security attention from U.S. naval forces in the future.

[23]See Doug Struck, "Dispute over Northwest Passage Revived: U.S. Asserts Free Use by All Ships, Canada Claims Jurisdiction," *Washington Post,* November 6, 2006.

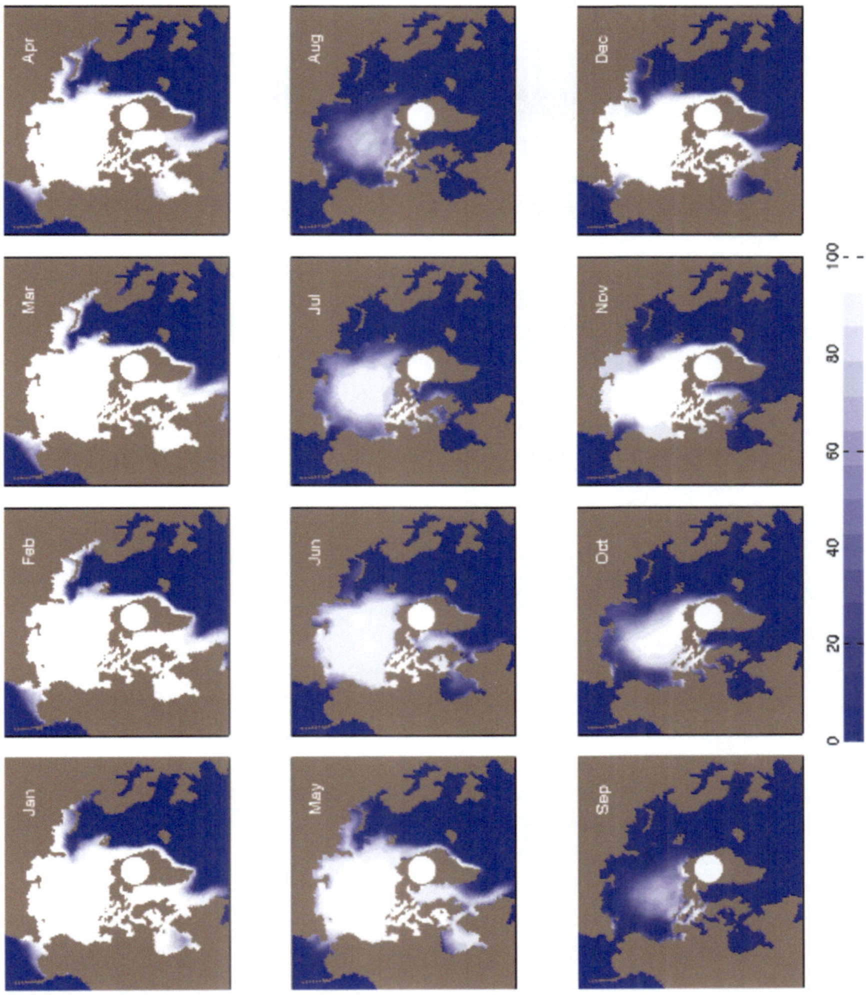

FIGURE 2.3 Possible sea-ice concentration in 2030, by month. The projection is a seven-member ensemble average from the Community Climate System Model Version 3 (CCSM3). A 50 percent ice concentration could mean that one out of two days will be sea-ice free or that on a given day the cover is 50 percent sea ice and 50 percent open water. The greenhouse gas scenario is from the Special Report on Emissions Scenarios (SRES) A1B, the moderate scenario used by climate models for the *Fourth Assessment Report (AR4) of the Intergovernmental Panel on Climate Change (IPCC)*. In the 20-year time frame, the greenhouse gas scenario has only a moderate bearing on the results.

International focus on the Arctic, including military and naval activities, has increased considerably over the past 3 years. A March 2009 Arctic policy statement from the Russian Security Council highlighted Russia's continued interests in the Arctic, including the need for maintaining combat potential and special Arctic military formations.[24] Russian military activities, such as resumption of Bear bomber flights, have indicated an increasingly assertive Russian posture. Canada, a strong U.S. ally, is preparing and planning to build six armed, ice-strengthened patrol vessels for Arctic sovereignty operations, establish a high-latitude logistics base, and construct a high-Arctic training facility.[25] Canadian naval forces are exercised each summer in Operation Nanook in northern Baffin Bay, with increasing U.S. participation. Norway, also an ally and NATO member, has moved its armed forces operations center to a more northern area of the country, and it utilizes aircraft and ice-strengthened vessels to actively patrol its Arctic waters. A Danish 2010-2104 defense plan envisions an Arctic military command and task force.[26] China and Korea, while not Arctic nations, have signaled their interest and intent to participate in the Arctic, including routine deployments of an icebreaking research vessel and a physical presence on the ground at Svalbard.

Key Arctic Operational Challenges

All of these developments in the Arctic have ramifications for future operations of U.S. naval forces. The Navy has many years of Cold War operating experience in the Arctic Ocean and sub-Arctic seas with submarines, and Marines trained regularly for deployment in northern Norway until the early 1990s. Surface and air operations have not been a priority for the Navy in the high latitudes for almost 25 years; so, today's naval forces lack experience and procedures for the challenges of these northern environments.

However, the demand for Coast Guard missions is evident and increasing. NSPD-66/HSPD-25's discussion of U.S. national security interests includes maritime presence and maritime security operations, homeland security, asserting a more active and influential presence, and exercising control over the U.S. EEZ, the continental shelf, and the contiguous zone.[27] These policy statements speak directly to Coast Guard responsibilities, reflecting aspects of the Coast Guard's 11 statutory missions. Since 2007, the Coast Guard has surged cutters, aircraft, boats, and special detachments to Arctic Alaska during the summer season to

[24]Katarzyna Zysk. 2010. "Russia's Arctic Strategy," *Joint Forces Quarterly,* Issue 57.

[25]Ross Graham, Director General Defence Research and Development Canada, Center for Operational Research and Analysis, "Impact of Climate Change on Canadian Naval Operations in the Arctic," presentation to the committee, February 4, 2010, Washington, D.C.

[26]BBC News, "Denmark Plans Forces for the Arctic," July 16, 2009; available at http://news.bbc.co.uk/2/hi/8154181.stm. Accessed July 28, 2010.

[27]See National Security Presidential Directive-66, Article III B 1; available at http://www.fas.org/irp/offdocs/nspd/nspd-66.htm. Accessed July 28, 2010.

increase competencies, develop Arctic operating procedures, and evaluate asset capabilities.

Unclassified national intelligence assessments suggest a low likelihood of significant conflict in the Arctic region in the foreseeable future.[28] Nevertheless, as a hedge against a more extreme scenario, the committee believes that as access to Arctic and sub-Arctic seas increases, the U.S. Navy and Coast Guard must be prepared for the potential requirement to exercise the full range of their capabilities in the Arctic. Establishing and maintaining U.S. naval capabilities in the Arctic will require attention to shore-based infrastructure, communications capabilities, competencies and operating experience, icebreakers and ice-capable ships, and combatant command issues.

Shore-Based Infrastructure

The Arctic encompasses vast areas with long distances between outposts. An infrastructure capable of supporting surface and air operations is sparse, particularly in Alaska and the western Arctic waters. This shortcoming especially affects the Coast Guard and its ability to execute mission responsibilities in the Bering, Chukchi, and Beaufort Seas. The distance from the Coast Guard's principal Alaskan base in Kodiak (a World War II–era naval air station) to the North Slope is approximately 800 nautical miles. The Coast Guard has used the concept of a "forward operating location," such as an Army National Guard hangar in Nome and commercial air facilities in remote communities, to support aircraft operations. But it is clear that for routine Arctic operations, the Coast Guard will have to develop a more robust methodology of supporting deployed assets. Given the high cost of constructing shore-side infrastructure and the reality that conventional piers are not feasible due to ice movement, using the inherent capabilities of a polar icebreaker as a mobile, multimission platform may be an attractive alternative. A major review of Coast Guard requirements now under way will address support infrastructure as well as better inform the need for protecting U.S. national security interests in the Alaskan Arctic.[29]

---

[28]As discussed in Chapter 4, the committee's independent evaluation, based on direct discussion with representatives from select Arctic nations and the April 2010 recent settlement of a long-standing territorial dispute between Russia and Norway, supports these assessments. See National Intelligence Council, *2025 Global Trends Report,* November 2008, p. 53; available at http://www.dni.gov/nic/PDF_2025/2025_Global _Trends_Final_Report.pdf. Accessed November 24, 2009. This unclassified report states, in part: "Although serious near-term tension could result in small-scale confrontation over contested claims, the Arctic is unlikely to spawn major armed conflict. Circumpolar states have other major ports on other bodies of water, so the Arctic does not pose any lifeblood blockade dangers. Additionally, these states share a common interest in regulating access to the Arctic by hostile powers, states of concern or dangerous non-state actors; and by their shared need for assistance from high-tech companies to exploit the Arctic's resources."

[29]The U.S. Coast Guard has commissioned a study—the *U.S. Coast Guard High Latitude Region Mission Analysis,* anticipated to be completed by the fourth quarter of 2010—to better define its

In the eastern Arctic (Baffin Bay plus the Greenland, Norwegian, and Barents seas, etc.), U.S. naval forces will clearly depend on allied nations for necessary shore-side support. The United States is well served by the time-tested NATO alliance, a history of operating with Canada, Norway, and Denmark, and established bases in the area (e.g., Thule in northeastern Greenland and Keflavik in Iceland). See Chapter 4 for a more complete discussion of allied forces.

Command, Control, Communications, Computers, Intelligence, Surveillance, and Reconnaissance Infrastructure

Effective operations by naval forces in the high latitudes will require improvements in command, control, communications, computers, intelligence, surveillance, and reconnaissance (C4ISR) capabilities for these areas. The robust set of geosynchronous Earth orbit (GEO) satellites provides reliable communications for latitudes between 65°N and 65°S. High-data-rate satellite communications are sparse over the polar regions. However, commercial low-rate service is available. Additionally, Global Positioning System (GPS) constellation coverage is not optimized for polar regions, and so its accuracy is reduced, but it still provides adequate performance for surface navigation. The committee believes that particular attention to the enhancement of satellite communications is vital because the requirements will become more compelling as Arctic operations increase.[30] See Chapter 5 for a discussion of C4ISR technical issues.

High-Latitude Competencies and Experience

The lack of operating experience by naval surface and air forces in cold-weather environments has resulted in a generation of naval personnel unfamiliar with the demands of operating in far-north areas, both at sea and ashore. The submarine community has generally avoided this geographic operations experience gap, although Arctic Ocean submarine operations have recently been conducted at a lower tempo than during the height of the Cold War. When the Coast Guard began to deploy air and non-icebreaker surface ships to the North Slope in 2007, the lack of practical operating experience was noteworthy. For example, the U.S. Coast Guard reported that it encountered a variety of challenges with operating the 25-ft Defender-class boat and the MH-65 helicopter from the temporary forward operating location during the 2007 North Slope training exercise. The challenges include a lack of communications networks, which limited the range

---

needs for routine operations in the Arctic region in support of NSPD-66/HSPD-25. Source: ADM Thad Allen, Commandant, U.S. Coast Guard, discussion with the committee, November 20, 2009, Washington, D.C.

[30]Related to this, the committee reviewed information on national imaging capabilities that may become increasingly important as Arctic activities increase. Information on national imaging capabilities and the Global Fiducials Library is available at http://gfl.usgs.gov/. Accessed April 12, 2010.

of operations to 60 miles. The unpredictability of sea ice and the prevailing sea state in the U.S. Arctic render the Coast Guard's current portfolio of small boats ineffective for safe operations.[31]

As naval forces confront the possibility of future operational requirements in Arctic and sub-Arctic areas, it is clear that a base of operating experience and competencies must be established. For Navy fleet surface and air assets, this could best be accomplished by exercises in open-water northern extensions of the North Atlantic. The opportunity to exercise with the forces of other NATO (or even non-NATO) nations is attractive. For example, the Canadian forces' annual Operation Nanook provides a venue for U.S. fleet forces to continue to build allied partnerships and gain Arctic operational experience. The Coast Guard should continue, at increasing tempos, the deployment of assets in Arctic Alaska and should extend these operations beyond the summer season to the degree possible.

In addition to Navy and Coast Guard operations in the Arctic, the Marine Corps should consider returning to northern engagement with allied partners as the current operations tempo in Central Command begins to diminish. The initial goal would be to develop a training program for individual Marines and small units whereby they are capable of surviving and sustaining themselves in the Arctic. Following the establishment of the training program, consideration should be given to potentially embarking small units with Navy and Coast Guard ships as they deploy in the Arctic, to conduct low-level exercises with allied forces. As examples, a small Marine unit might be inserted ashore from a Coast Guard icebreaker in an ice-covered waterway on a search and rescue mission; or an amphibious ship with landing crafts and Marines might conduct HA/DR or noncombat security exercises with Canadian forces.

Polar Icebreakers and Ice-Capable Combatant Ships

The Navy has no surface combatants hardened for ice operations. Additionally, a recent report by the National Research Council highlighted the fact that two of the nation's three multimission polar icebreakers are at the end of their designed service lives and that the icebreaker operating budgets are controlled by the National Science Foundation.[32] Considering projected increases in resource

---

[31]See *Report to Congress: U.S. Coast Guard Polar Operations,* 2008, p. 12. Available at http://www.uscg.mil/hq/cg5/cg513/docs/FY08_OMNIBUS_Polar_Ops_Report.pdf. Accessed June 4, 2010.

[32]The three U.S. Coast Guard icebreakers are the *Polar Star,* commissioned into service in 1976; the *Polar Sea,* commissioned in 1978; and the *Healy,* commissioned in 2000. Each vessel was designed for a 30-year service life. The *Polar Star* has been in caretaker status since 2006. The *Polar Sea* is in operational condition but, because of its age, requires increasing amounts of maintenance to remain in operation. See National Research Council, 2007, *Polar Icebreakers in a Changing World: An Assessment of U.S. Needs,* The National Academies Press, Washington, D.C. See also Ronald O'Rourke, 2009, *Coast Guard Polar Icebreaker Modernization: Background, Issues, and Options for Congress,* CRS 7-5700, RL34391, Congressional Research Service, Washington, D.C., May 29.

development, maritime transportation, and international competition in the Arctic, U.S. icebreaking resources are clearly inadequate to meet national needs. This deficiency is particularly significant given the recent and continuing investment in icebreaking resources by other countries, including China, Russia, Japan, South Korea, and the European Union (see Table 2.1 and Box 2.1).[33]

Icebreakers provide important naval force capabilities. These ships permit year-round access to Arctic waters, and their design includes high endurance and lengthy on-station patrol times. They can escort ice-strengthened vessels, hangar and support helicopters, carry cargo, accommodate embarked detachments, conduct scientific research, refuel and re-provision other vessels, and provide contingency command, control, and communications services. Polar icebreakers may also be an economical alternative for executing Coast Guard missions in the ice-influenced EEZ around Alaska. In general, icebreakers provide the nation with its only sovereign surface presence in the Arctic, complementing U.S. submarine and air capabilities.

Unfortunately, the Coast Guard has found that maintaining a core fleet of icebreaking capability has been challenging. Construction of the newest of the three icebreakers required 25 years of studies and budget requests before USCGC *Healy* was commissioned in 2000. Due to waning Cold War requirements and the use of icebreakers in supporting research, the Office of Management and Budget transferred the icebreaker operating budgets to the National Science Foundation in 2005. As the challenges of a transforming Arctic grew more apparent, the Commandant of the Coast Guard argued strongly for recapitalization of the older icebreakers, an initiative that has been supported by a 2007 NRC report, the Joint Chiefs of Staff, and NSPD-66/HSPD-25. The Coast Guard's Acquisition Directorate has commissioned a polar icebreaker business-case analysis, reportedly to be completed in late 2010, to evaluate recapitalization alternatives as the foundation for a possible acquisition project.

Related to this, the Navy should evaluate requirements for future surface combatants and auxiliaries to operate in ice-covered waters. A recent report by the Center for Naval Analyses noted that current surface combatants might be modified or retrofitted for Arctic operations by having steel added around the

---

[33]For example, a 2007 National Research Council report that lists a world inventory estimate of polar and Baltic icebreakers states that Russia has by far the largest fleet of icebreakers, although some of them are aging and some are used to keep supply lines open to Russia's Arctic coastal settlements. Data in the 2007 study indicate that Russia has 18 icebreakers, 7 of which are nuclear powered; Finland and Sweden are reported to have 7 icebreakers each; and Canada is reported to have 6 icebreakers. See National Research Council, 2007, *Polar Icebreakers in a Changing World: An Assessment of U.S. Needs,* The National Academies Press, Washington, D.C., pp. 57-59. China, Japan, and South Korea have also made recent investments in new icebreakers targeted for polar research. For example, see "China to Build Own Icebreakers for Poles"; available at http://www.shanghaidaily.com/sp/article/2009/200910/20091008/article_415716.htm. Accessed November 24, 2009.

TABLE 2.1 Global Polar Icebreaker Inventory by Country[a]

Country of Ownership	Total Icebreakers	Inventory by Age (Decade of Entering Service)	Propulsion[b]
Russia	18	5—1970s	7—N
		8—1980s	7—DE
		3—1990s	4—D
		2—2000s	
Finland	7	2—1970s	7—DE
		2—1980s	
		3—1990s	
Sweden	7	3—1970s	6—DE
		1—1980s	1—D
		3—2000s	
Canada	6	1—1970s	5—DE
		3—1980s	1—D
		1—1990s	
		1—2000s	
Netherlands	3	1—1970s	1—DE
		2—1980s	2—D
United States[c,d]	3	2—1970s	3—DE
		1—2000s	

[a]In addition to the inventory listed in this table, the following countries own and operate at least one operational icebreaker: Argentina, Australia, China, Germany, Japan, and Norway. China, Japan, and South Korea are also reportedly investing in additional icebreaker capacity for polar research.

[b]N = nuclear, DE = diesel electric, and D = geared diesel.

[c]The *Nathaniel B. Palmer,* commissioned in 1992, is a 308-ft-long, geared diesel vessel, chartered and operated by the U.S. National Science Foundation. The *Palmer* has limited icebreaking capability (3 feet thick at speeds of 3 knots) and is used exclusively as a research vessel in the Antarctic. As a single-mission research vessel with limited icebreaking capability, it is considered by many to be more of an oceanographic research ship than a true icebreaker. The *Palmer* is not included in these numbers.

[d]On June 25, 2010, the U.S. Coast Guard announced that its only operational heavy icebreaker, the *Polar Sea,* suffered an unexpected engine casualty and will be unable to deploy on its scheduled fall 2010 Arctic patrol. *Polar Sea* will likely be in a maintenance status and unavailable for operation until at least January 2011. *Polar Sea* was commissioned into service in 1978 with a 30-year service life. In 2006 the Coast Guard completed a rehabilitation project that extended its service life to 2014. *Polar Star*, the Coast Guard's other heavy icebreaker, commissioned in 1976, is currently in the process of being reactivated but will not be operational for deployment until 2013. The *Polar Star* was placed in a caretaker status in 2006. Currently, the 420-foot USCGC *Healy,* commissioned in 2000, is the service's sole operational polar region icebreaker. While the *Healy* is capable of supporting a wide range of Coast Guard missions in the polar regions, it is a medium icebreaker capable of breaking ice up to 4.5-feet thick at three knots. (USCG *Compass*, June 25, 2010, and USCGC *Healy* website).

SOURCE: Derived from Arctic Marine Shipping Assessment Data Base, National Research Council, 2007, *Polar Icebreakers in a Changing World: An Assessment of U.S. Needs,* The National Academies Press, Washington, D.C.; and Ronald O'Rourke, 2009, *Coast Guard Polar Icebreaker Modernization: Background, Issues, and Options for Congress,* CRS 7-5700, RL34391, Congressional Research Service, Washington, D.C., May 29.

BOX 2.1
Conclusions and Recommendations from 2007 NRC Report on U.S. Coast Guard Polar Icebreaker Needs

The following conclusions and recommendations are reprinted from National Research Council, 2007, *Polar Icebreakers in a Changing World: An Assessment of U.S. Needs,* The National Academies Press, Washington, D.C., pp. 2-3:

> [B]oth operations and maintenance of the polar icebreaker fleet have been underfunded for many years, and the capabilities of the nation's icebreaking fleet have diminished substantially. Deferred long-term maintenance and failure to execute a plan for replacement or refurbishment of the nation's icebreaking ships have placed national interests in the polar regions at risk. The nation needs the capability to operate in both polar regions reliably and at will. Specifically, the [2007] committee recommends the following:
>
> - The United States should continue to project an active and influential presence in the Arctic to support its interests. This requires U.S. government polar icebreaking capability to ensure year-round access throughout the region. . . .
> - The United States should maintain leadership in polar research. This requires icebreaking capability to provide access to the deep Arctic and the ice-covered waters of the Antarctic.
> - National interests in the polar regions require that the United States immediately program, budget, design, and construct two new polar icebreakers to be operated by the U.S. Coast Guard.
> - To provide continuity of U.S. icebreaking capabilities, the POLAR SEA should remain mission capable and the POLAR STAR should remain available for reactivation until the new polar icebreakers enter service.
> - The U.S. Coast Guard should be provided sufficient operations and maintenance budget to support an increased, regular, and influential presence in the Arctic. Other agencies should reimburse incremental costs associated with directed mission tasking.
> - Polar icebreakers are essential instruments of U.S. national policy in the changing polar regions. To ensure adequate national icebreaking capability into the future, a Presidential Decision Directive should be issued to clearly align agency responsibilities and budgetary authorities.

waterline, but that this would provide only marginal capability.[34] Effective vessel operation in sea ice, even in ice concentrations less than 10 percent, requires not only hull protection but also strengthened and upgraded propellers, rudders, seawater intakes, and hull-mounted sensors. In this committee's opinion, it is better to build ice-capable ships from the keel up, either by incorporating the capability into current designs or by designing a new class of vessels, as other nations have chosen to do. For example, the Canadian Navy is designing a class of armed, ice-capable patrol vessels.[35] As future U.S. Navy surface ships' needs are evaluated, existing ship classes modified and upgraded for operating in sea ice may offer the right level of naval capability. It may also be wise to build more robust under-ice capability into some fraction of future Virginia-class nuclear-powered attack submarines to support the projected increase in under-ice missions.[36]

FINDING 2.3: The nation has very limited icebreaker capability, which could limit the U.S. ability to train, operate, and engage in the Arctic. Furthermore, as noted in a 2007 National Research Council report, "both operations and maintenance of [the] polar icebreaker fleet have been underfunded for many years, and the capabilities of the nation's icebreaking fleet have diminished substantially" and, among other things, "the U.S. Coast Guard [USCG] should be provided sufficient operations and maintenance budget[s] to support an increased, regular, and influential presence in the Arctic."[37] Moreover, U.S. national icebreaker assets are old, obsolete, and under the control of another agency that does not have a national security operational mandate. The present committee believes that future USCG missions in the Arctic will require autonomy and command of their vessels.

RECOMMENDATION 2.3: In order to support the U.S. naval forces' missions in the Arctic, the U.S. Coast Guard (USCG) needs icebreaker capabilities under

---

[34]Michael D. Bowes. 2009. *Impact of Climate Change on Naval Operations in the Arctic,* CAB D0020034.A3/1REV, Center for Naval Analyses, Alexandria, Va., April.

[35]These vessels are reported to be 110 meters in length, 6,900 tons, and capable of operating in first-year ice with old-ice inclusions.

[36]Public news articles have reported that the nuclear-powered submarine *Texas* (SSN-775) and its 134-member crew recently completed an Arctic mission. The *Texas* reportedly broke through the ice near the North Pole and stayed on the surface for 24 hours and was the third U.S. submarine to visit the region in 2009. For deployment on Arctic missions, Virginia-class attack submarines such as the *Texas* reportedly carry an "Arctic sensor suite" similar to that carried by the older Los Angeles-class submarines that have previously traversed waters near the North Pole. This sensor suite is not a built-in capability, but instead only an add-on before deploying to an Arctic region. A Navy spokesperson has been quoted as saying that "Virginia-class submarines are not ice-hardened, and there are no plans to add ice-hardening to their designs." See Rick Rozoff, "Loose Cannon and Nuclear Submarines: West Prepares for Arctic Warfare," *CanWest News Service,* November 16, 2009; and Zachary M. Peterson, "VA-Class Submarines Carry Arctic Sensor Suite in Northern Waters," *Inside the Navy,* November 30, 2009.

[37]National Research Council. 2007. *Polar Icebreakers in a Changing World: An Assessment of U.S. Needs,* The National Academies Press, Washington, D.C., p. 102.

its operational control. While there are other national requirements for such ships, action should be taken to provide these operational capabilities to the USCG. Therefore, the Chief of Naval Operations should support the initiatives of the Commandant of the Coast Guard to define future USCG icebreaker needs. As such, future U.S. national icebreaker assets should be defined as part of a holistic force structure that also accommodates ongoing National Science Foundation-sponsored polar research needs.

Arctic Command Issues

As world conditions and defense needs evolve, the Unified Command Plan (UCP) is occasionally updated. For example, in 2004 U.S. COCOM responsibilities in the Middle East were realigned. The more recent creation of Africa Command was a major realignment of COCOM responsibilities. The committee believes that it is time to review the Arctic in this regard.

The Arctic contains areas of responsibility of three combatant commanders—Commander, U.S. European Command (EUCOM); Commander, U.S. Northern Command (NORTHCOM); and Commander, U.S. Pacific Command (PACOM). Because Arctic boundaries are so complex and are subject to change, the U.S. command structure for this region should be reviewed. Many international and interagency issues are in play in the Arctic, including search and rescue, navigation, and environmental rules related to operations at sea. The primary issues in the Arctic at the current time are legal as opposed to military in nature, so a review of the U.S. command structure should involve the State Department, Coast Guard, and perhaps other agencies as well. Because defense posture, particularly in the Arctic, is now more focused on engagement rather than on military force, DOD combatant command authority structure for the Arctic region should be as consistent as possible with State Department areas of responsibility.

FINDING 2.4: The current situation of three combatant commanders—Commander, U.S. European Command; Commander, U.S. Northern Command; and Commander, U.S. Pacific Command—having overlapping areas of responsibility for the Arctic was perhaps workable when the Arctic was less important than it is rapidly becoming. This division of responsibility in the Arctic is inconsistent with U.S. national interests and does not match the command structure of other U.S. agencies (such as the Department of Homeland Security and the U.S. Department of State) in this increasingly significant region of the world.

RECOMMENDATION 2.4: The Chief of Naval Operations should engage the Joint Chiefs of Staff in a review of combatant commanders' responsibilities for the Arctic, with the goal of ensuring the most effective command structure. Interagency considerations, including but not limited to the U.S. Department of State, should be included in these deliberations.

MAINTAINING CAPABILITIES

As the operating environment changes, the Navy needs to understand how its ability to project power will be impacted. For example, regarding strike warfare, a 2003 study by the Center for Naval Analyses looked at the susceptibility of carrier flight deck personnel to heat stress.[38] This study was unrelated to climate change consideration and was driven by requirements at that time for prolonged carrier summer operations in the Arabian Gulf. The study found that under such high-temperature conditions (heat indices reaching 140°F on the flight deck), crew become fatigued more quickly than under normal conditions, and crew endurance became the limiting factor in the ability of the airwing to maintain high-tempo operations. The study estimated that the firepower potential of the airwing (sorties per hour) was reduced to about two-thirds of that possible in temperate climates.

As stated in the Navy's Climate Change Roadmap,[39] one of the Navy's priorities is to ensure that it is fully mission capable as climate changes. Because virtually all Navy operations are subject to the effects of weather, climate change could prove challenging. If severe weather becomes more frequent as climate changes, training and readiness can also be affected. Although there is not too much that the Navy can do to prevent this or create more weather-resilient platforms, it will be increasingly important for the Navy to ensure a robust weather monitoring and prediction capability. These are critical capabilities now and will, perhaps, become even more critical in the future.

FINDING 2.5: In the post–Cold War era, the U.S. Navy has had a very limited surface ship presence in true northern latitude, cold-weather conditions. According to information presented to the committee, the U.S. military as a whole has lost most of its competence in cold-weather operations for high-Arctic warfare.

RECOMMENDATION 2.5: The Chief of Naval Operations, the Commandant of the Marine Corps, and the Commandant of the Coast Guard should establish a strong and consistently funded effort to increase Arctic operations and share lessons, including with allies. In the immediate term, the Navy should begin Arctic training and the Marine Corps should also reestablish a cold-weather training program.

---

[38] Angelyn Jewell, Timothy Roberts, and Timothy DeBisschop. 2003. *Susceptibility of Carrier Flight Deck Crewmen to Heat Stress,* Center for Naval Analyses, CRM D0008026.A2/Final, Alexandria, Va., March.

[39] Task Force Climate Change, Oceanographer of the Navy. 2010. *U.S. Navy Climate Change Roadmap,* Washington, D.C., April.

HEALTH, DISEASE, AND CLIMATE

During its deliberations, the committee was briefed on the significant implications of climate change on health.[40] These changes may also have an impact on global hot spots or cause concern for U.S. naval forces. The implications are from both the primary effects of changes in patterns and intensity of disease and the secondary effects of disease on populations already stressed by malnutrition and a burden of chronic disease. An example is a situation where a chronically undernourished infant becomes afflicted with acute diarrheal disease. The threat is immediate from the acute disease and long term due to the known deleterious effects of disease on growth and development.

The spectrum of climate change impacts on human health and disease is wide, including the European heat wave of 2003, which took over 30,000 lives, and extreme precipitation events leading to outbreaks of disease—such as the diarrheal disease outbreak caused by *Cryptosporidium* in Milwaukee, Wisconsin, following heavy spring rains in 1993.[41] On the other hand, drought can be just as harmful, because it leads to a diminished and more likely contaminated water supply resulting in outbreaks of diseases such as cholera and the spread of disease such as Rift Valley Fever in Africa, or the hantavirus. There are also several cited examples suggesting that climate change has resulted in the introduction of certain infectious diseases into previously unaffected areas. Examples include outbreaks of malaria in the highlands of East Africa, the spread of Dengue fever into Mexico and most likely soon the United States, and the discovery for the first time of two known pathogens—cryptococcosis and *Vibrio vulnificus*—in the Pacific Northwest and Vancouver.[42] (As an example of recent regional climate

---

[40]RADM Ali S. Khan, MD, MPH (U.S. Public Health Service), Assistant Surgeon General and Deputy Director (acting), National Center for Emerging and Zoonotic Infectious Diseases, Department of Health and Human Services/Centers for Disease Control and Prevention, "Climate and Health: Preparing for and Communicating Complexity"; and Eileen Choffnes, Scholar/Director, Forum on Microbial Threats, The National Academies Institute of Medicine, "Ecological, Environmental, and Infectious Disease Implications of Global Climate Change and Extreme Weather Events"—presentations to the committee, February 4, 2010, Washington, D.C.

[41]See, for example, "The 2003 Heat Wave in Europe: A Shape of Things to Come? An Analysis Based on Swiss Climatological Data and Model Simulations," *Geophysical Research Letters,* Vol. 31, 2004; available at http://www.agu.org/pubs/crossref/2004/2003GL018857.shtml. See also "A Massive Outbreak in Milwaukee of Cryptosporidium Infection Transmitted Through the Public Water Supply," *New England Journal of Medicine,* July, Vol. 331, No. 3, pp. 161-167, 1994.

[42]Cryptococcosis is a potentially fatal fungus disease. This fungus is ordinarily found in soil. It is the cause of the most common life-threatening meningitis in AIDS patients. Early in the epidemic, approximately 5 to 8 percent of patients with AIDS developed cryptococcal infection. Cryptococcosis mainly occurs in the tissues covering the brain, spinal cord in the lungs, and on the skin. *Vibrio vulnificus* is a gram-negative bacillus that only affects humans and other primates. It is in the same family as bacteria that cause cholera. The first documented case of disease caused by the organism was in 1979. *V vulnificus* is usually found in warm, shallow, coastal salt water in temperate climates throughout most of the world.

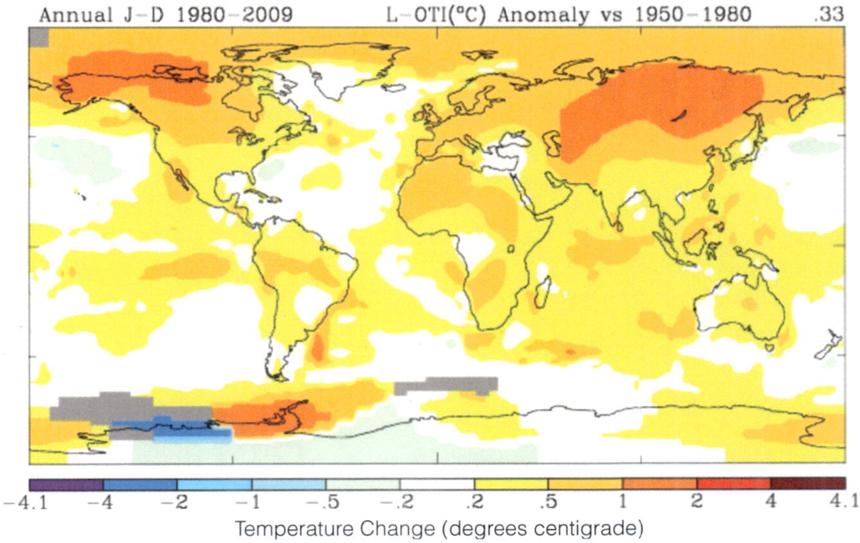

FIGURE 2.4 Regional variation in global temperature trends over the past 30 years. Yellow, orange, and red designate average increase in temperate (°C) from 1980 through 2009, compared with the previous three decades. Warming has been the greatest in the Northern Hemisphere. SOURCE: NASA Goddard Institute for Space Studies, http://data.giss.nasa.gov/gistemp/.

change differences, Figure 2.4 provides a view of variation in regional temperature change over the past 30 years.)

Both the National Research Council and the World Health Organization (WHO) have released reports regarding the estimated effects of climate change on disease vectors and human health.[43] According to these reports, the impacts of climate on human health will not be evenly distributed around the world. Developing countries, particularly in small island states, arid and high mountain zones, and densely populated coastal areas, are considered to be particularly vulnerable to these impacts. Based on these studies, human health and disease effects may exacerbate climate change impacts in certain regions of the globe, impacting the United States and its allies, and they are also cause for U.S. naval forces to consider new-disease vectors when preparing to respond to new missions.

[43]See National Research Council, 2001, *Under the Weather: Climate, Ecosystems, and Infectious Disease,* 2001, National Academy Press, Washington, D.C. See also World Health Organization, 2009, *Climate Change and Health, Report by the Secretariat,* Geneva, March.

FINDING 2.6: Climate change is impacting the geographic distribution of disease and, in many instances, its intensity. As disease vectors change their distribution, the result is larger populations at risk. In addition, previously unexposed populations may be more severely affected, particularly when they carry the burdens of malnutrition and chronic disease.

RECOMMENDATION 2.6: U.S. naval leadership should consider the impact of changing disease vectors on the population when forecasting the impact of climate change, and should also consider climate-change-related changing disease vectors in preparing troops for response to missions around the globe.

3

Infrastructure Issues

NAVAL INFRASTRUCTURE AND GLOBAL AND LOCAL SEA-LEVEL RISE—BACKGROUND

Among the many manifestations of climate change projected for the next several decades, sea-level rise is both highly certain to occur and highly certain to come with economic costs. This chapter begins with a review of historic sea-level change and the lack of precision in future sea-level prediction for future naval planning. The chapter then discusses the regional aspects of sea-level rise and the need for individual naval base assessments—to include the role of storm surge. The chapter concludes by reviewing recent preliminary naval coastal installation vulnerability assessments and makes suggestions for improving these assessments going forward.

The rate of sea-level rise has apparently accelerated in recent decades, although in the projections of further rise during the current century, an additional large uncertainty is how global change will be manifested locally. Local impact will in large part depend on changes in ocean circulation. An additional exacerbating condition relates to storm surge and the likelihood that some storms will intensify with further warming of the atmosphere and ocean.[1] Sea-level rise is not uniform around the globe, and the potential coastal impact of regional sea-level rise is not linear with elevation. Sea-level rise even in the lower range of projections will challenge the utility and perhaps even viability of some shore-based

[1] Morris A. Bender, Thomas R. Knutson, Robert E. Tuleya, Joseph J. Sirutis, Gabriel A. Vecchi, Stephen T. Garner, and Isaac M. Held. 2010. "Modeled Impact of Anthropogenic Warming on the Frequency of Intense Atlantic Hurricanes," *Science*, Vol. 327, No. 5964, pp. 454-458.

facilities. (Figure 3.1 and Box 3.1 provide illustrations of potential sea-level rise impact and the compounding impact of sea-level rise and storm surge.)

Precision in the measurement of changes in globally averaged sea level was improved substantially in the early 1990s with the deployment of the TOPEX/

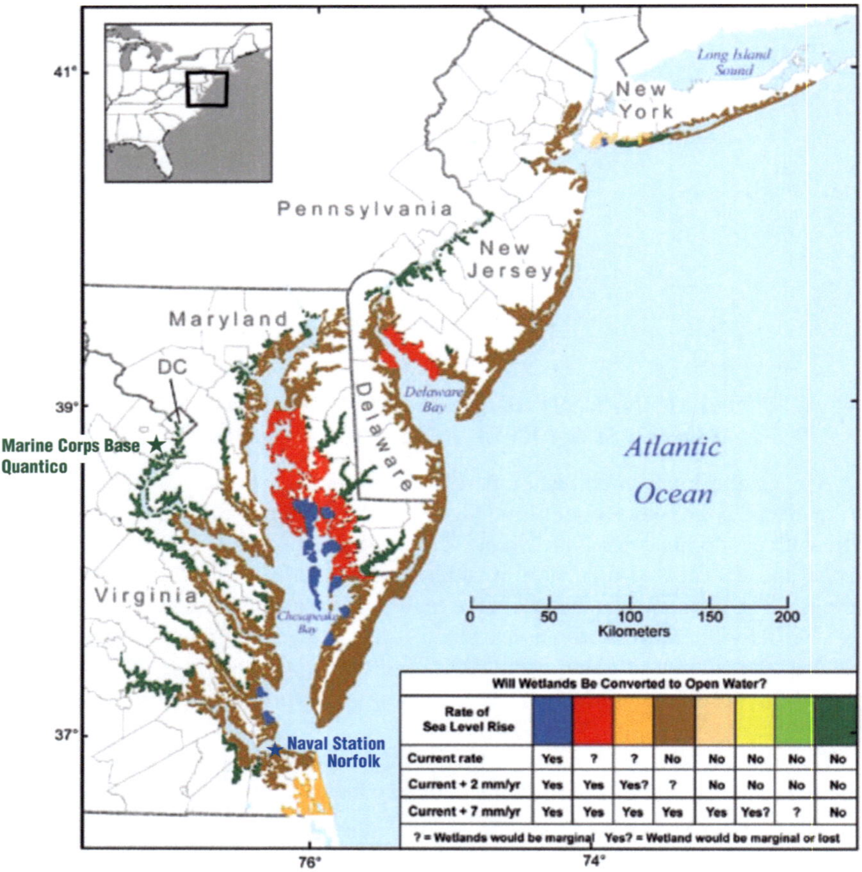

FIGURE 3.1 Potential regional impact of future sea-level rise. Several static and dynamic models are being developed for projecting the regional impact of sea-level rise. This figure shows potential impact to wetlands in the U.S. mid-Atlantic region under various sea-level rise scenarios (areas where wetlands would be marginal or lost [i.e., converted to open water] under three sea-level rise scenarios, in millimeters [mm] per year [yr]). Such scenarios may be applicable on a gross scale for judging first-order impact on naval installations. SOURCE: Reprinted from Figure ES.2, *Coastal Sensitivity to Sea-Level Rise: A Focus on the Mid-Atlantic Region*. A report by the U.S. Climate Change Science Program and the Subcommittee on Global Change Research, 2009, U.S. Environmental Protection Agency, Washington, D.C.

Poseidon satellite altimeter, followed by later high-precision satellite altimeter missions. As a result, it is now possible to detect acceleration in sea-level rise over the past few decades.[2] The current estimated rate (3 mm/year) is already at the upper limit of the range of global sea-level rise projections that were presented 20 years ago in the first Intergovernmental Panel on Climate Change (IPCC) assessment. Although this rate is small relative to the magnitude of tidal excursions at most localities, the probability of sea level rising at this rate or faster (while adding to tidal excursions and storm surges) over the next century requires serious assessment of the implications for coastal facilities.

Sea level has changed in the past with major climate cycles, dramatically so, relative to projections for future decades. During the last glacial maximum about 18,000 years ago, the accumulation of ice on land resulted in the surface ocean being approximately 120 meters below current levels.[3] With the melting of most of this ice over a few thousand years, sea level rose to near its current level. As the human population grew over the past few centuries, a relatively stable sea level led to the extensive development of coastal margins for habitation and economic enterprise. This legacy of successful coastal development reflects the history of sea level within tidal variation and the degree to which related vulnerability issues are occasionally amplified by coastal storms. A 1900 hurricane in Galveston, Texas, with 8,000 fatalities, is a case in point. A rise of 0.5 meter over several decades will be of enormous consequence for coastal cities and naval installations that experience only small natural tidal variation. For example, such an increase in sea level will have significant consequences for St. Petersburg, Florida (typical tidal excursion up to 0.7 meter), but virtually none for Anchorage, Alaska (typical tidal excursion 10 meters).

The 2007 IPCC *Fourth Assessment Report* (AR4) projected changes in global average sea level under various scenarios considering glacier and ice sheet mass loss, but recognized that there was much uncertainty in the results. This lack of precision is because observational records of sea-level rise are short and therefore subject to uncertainty; in addition, the understanding of how glaciers and ice sheets will respond to increased temperature changes is very poor.[4]

Work published since the 2007 IPCC report suggests that loss of ice from small ice bodies (e.g., mountain glaciers and small ice caps) may have been

---

[2]Martin Sommerkorn and Susan Joy Hassol (eds.). 2009. *Arctic Climate Feedbacks: Global Implications,* World Wildlife Fund, International Arctic Programme, Oslo, p. 13.

[3]The Last Glacial Maximum refers to the period in Earth's history when the glaciers were at their thickest and the sea levels at their lowest. See Peter U. Clarke, Arthur S. Dyke, Jeremy D. Shakun, Anders E. Carlson, Jorie Clark, Barbara Wohlfarth, Jerry X. Mitrovica, Steven W. Hostetler, and A. Marshall McCabe, 2009, "The Last Glacial Maximum," *Science,* August 7, Vol. 325, No. 5941, pp. 710-714.

[4]IPCC. 2007. *Fourth Assessment Report of the Intergovernmental Panel on Climate Change,* Cambridge University Press, United Kingdom and New York, Chapter 10.

BOX 3.1
Sea-Level Rise and Storm Surge

The effect of rising mean sea levels at naval coastal installations will be felt most profoundly during extreme storm conditions when strong winds and low atmospheric pressure bring about a temporary and localized increase in sea level known as a storm surge. Storm surge is simply water that is pushed toward the shore by the force of the winds swirling around the storm. This advancing surge combines with the normal tides to create the storm tide, which can increase the mean water level 15 feet or more. In addition, wind-driven waves are superimposed on the storm tide (see Figure 3.1.1).

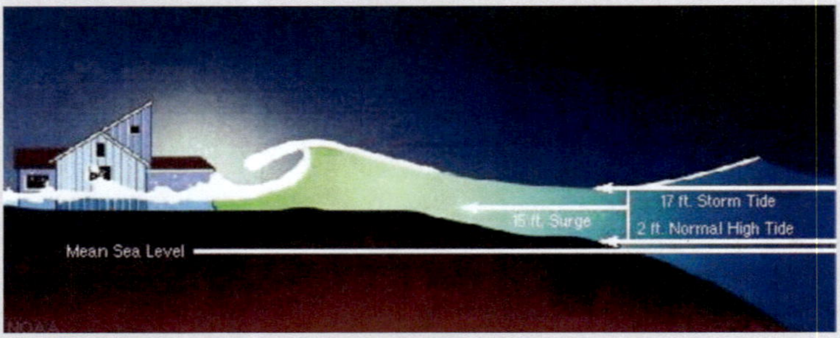

FIGURE 3.1.1 Sea-level rise and storm surge. SOURCE: Courtesy of NOAA.

This rise in water level can cause severe flooding in coastal areas, particularly when the storm surge coincides with the normal high tides. Storm surges occurring on higher mean sea levels will enable inundation and damaging waves to penetrate further inland, increasing flooding, erosion, and the subsequent detrimental impacts on built infrastructure and natural ecosystems. At coastal naval instal-

underestimated[5,6] and that major changes in Greenland and Antarctic ice sheet dynamics can take place over relatively short timescales. These changes should

[5]Mark F. Meier, Mark B. Dyurgerov, Ursula K. Rick, Shad O'Neel, W. Tad Pfeffer, Robert S. Anderson, Suzanne P. Anderson, and Andrey F. Glazovsky. 2007. "Glaciers Dominate Eustatic Sea-Level Rise in the 21st Century," *Science,* Vol. 317, pp. 1064-1067.

[6]David B. Bahr, Mark Dyurgerov, and Mark F. Meier. 2009. "Sea-Level Rise from Glaciers and Ice Caps: A Lower Bound," *Geophysical Research Letters,* Vol. 36, L03501.

lations where coastlines lie less than 10 feet above mean sea level, the impact from intense storm tides can be especially severe. For example, the August 2005 Hurricane Katrina—a Category 3 hurricane at landfall with 120 mph (195 km/h) sustained winds—produced a U.S. record storm surge, with peak surges estimated at 30 to 35 feet high (9 to 10 meters) and inland water reaching 6 to 12 miles (10 to 19 km) from the beach.

The impact of sea-level rise and the inundation caused by storm surge on naval facilities in a particular area is determined in part by the slope of the continental shelf and shoreline elevation. A shallow or gentle slope off the coast (right side of Figure 3.1.2) will allow a greater surge to inundate coastal communities. Communities with a steeper continental shelf (left side of Figure 3.1.2) will not see as much surge inundation, although large breaking waves can still present major problems.

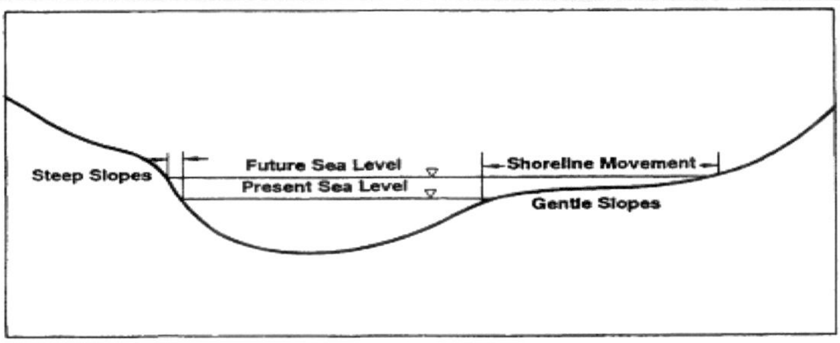

FIGURE 3.1.2 Continental shelf and shoreline elevation representation.

SOURCE: Adapted from data provided by NOAA.

be considered when planning for future sea-level increases and for naval scenarios in response. The mechanisms that could potentially drive rapid ice sheet change include meltwater drainage with an increase in basal lubrication, glacier surging due to basal hydrology changes, ice sheet/ocean interactions enhancing glacier melt by contact with warming seawater, and ice sheet thinning due to the loss of ice shelf buttressing.[7] It has also been recognized that discharge rather than surface mass loss alone can dominate ice sheet mass balance in Greenland; this implies the importance of ice dynamics in future sea-level projections.[8] Although ongoing studies are attempting to incorporate these and other processes into future ice sheet dynamic models, and hence sea-level rise projections, the state of knowledge to accomplish this is at present lacking.

Several studies since the last IPCC report have approached the prediction of future sea level by refining previous ice loss estimates using different approaches. By considering observational data from the accelerated loss of ice from small glaciers and the dynamics of small marine terminating glaciers, several researchers have suggested an additional 0.1 to 0.25 meter to the global sea-level estimates reported in the IPCC's Fourth Assessment.[9] The most recent work on this topic proposes that a lower bound for the global contribution of melt from small glaciers and ice caps could be as much as 0.37 meter over the next 100 years.[10] In 2007, the IPCC projections of the contribution from small glacier and ice cap melt to the sea level were between only 0.10 and 0.12 meter. In regard to the larger ice masses, recent work using a kinematic approach to determine plausible velocities of outlet glaciers in both Greenland and Antarctic has led to an indication of global average sea-level rise ranging from 0.4 meter to 2 meters by 2100.[11] An examination of the various estimates leads the committee to regard a 0.8 meter global average sea-level rise by 2100 as a reasonable planning target for naval leaders. These estimates are subject to change and should be reviewed routinely.

In 2008, the National Intelligence Council judged that more than 30 U.S. military installations were already facing elevated levels of risk from rising sea levels. Research is ongoing in this important area, but more is needed in order first to better quantify glacier dynamics in a warming world and then to assess the impact of these dynamics on glacier ice loss and future sea-level rise.

Regional Aspects of Sea-Level Rise

Although a great deal of attention has focused on the question of mean sea-level rise, it is the regional variations that are of most serious concern to naval

[7]Jonathan L. Bamber, Riccardo E.M. Riva, Bert L.A. Vermeersen, and Anne M. LeBrocq. 2009. "Reassessment of the Potential Sea-Level Rise from a Collapse of the West Antarctic Ice Sheet," *Science,* Vol. 324, pp. 901-903.

[8]E. Rignot, J.E. Box, E. Burgess, and E. Hanna. 2008. "Mass Balance of the Greenland Ice Sheet from 1958 to 2007," *Geophysical Research Letters,* Vol. 35, L20502.

[9]Mark F. Meier, Mark B. Dyurgerov, Ursula K. Rick, Shad O'Neel, W. Tad Pfeffer, Robert S. Anderson, Suzanne P. Anderson, and Andrey F. Glazovsky. 2007. "Glaciers Dominate Eustatic Sea-Level Rise in the 21st Century," *Science,* Vol. 317, pp. 1064-1067.

[10]David B. Bahr, Mark Dyurgerov, and Mark F. Meier. 2009. "Sea-Level Rise from Glaciers and Ice Caps: A Lower Bound," *Geophysical Research Letters,* Vol. 36, L03501.

[11]W. Tad Pfeffer, Joel T. Harper, and Shad O'Neal. 2008. "Kinematic Constraints on Glacier Contributions to 21st Century Sea Level Rise," *Science,* Vol. 321, pp. 1340-1343.

forces and their installations. Worst-case regional changes are more than an order of magnitude greater than the global mean (see Figures 3.1 and 3.2). Global mean sea-level rise is primarily of concern to scientists attempting to balance global heat and water budgets. It is notable that in some regions of the world local sea level has been falling for more than a decade, albeit most places are experiencing apparent rises.

The large inferred relative regional changes in sea level arise from the combined effects of local changes in the freshwater balance (evaporation, precipitation, runoff, groundwater drawdown, and subsurface percolation); changes in the heat exchange with the atmosphere; and—probably the dominant effect—shifts in the wind patterns (both locally and globally) and isostatic rebound. Understanding the trends in regional sea level requires an understanding of oceanic general circulation, based on in situ and remote sensed data, regional and global circulation models, and shifts in land-ice volume.

Recently, a number of groups[12,13] have begun making claims of skill in decadal forecasting of regional sea-level change. These, in turn, are forecasts of the ocean circulation that directly involve decadal forecast ability not just of the ocean but also of the overlying atmosphere and cryospheric components. The evaluation of such efforts, a determination of their probability of success, and the design of sensible naval responses to protect coastal assets present a major scientific problem for which the Navy requires its own expertise.

In many situations, neither the regional nor the global sea level is directly of primary interest. Rather it is the increased vulnerability of coastal areas to extreme events (storm surges) and the dependence of such events upon changes in regional relative sea level, tidal amplitudes, and the nature of extreme meteorological forces that are of greatest importance.[14,15] Therefore, evaluating future risks for naval installations involves an understanding of changes in storm frequency as well as local sea-level rise. Each and every naval facility has a unique configuration and requires ongoing evaluation of changing risks as the climate changes.

Figure 3.2 shows the estimated trend in sea level from use of altimetric data alone.[16] Figure 3.3 displays the anomaly (the global mean having been subtracted) from the combination of altimetry with temperature, salinity, and transport esti-

[12]Jianjun Yin, Michael E. Schlesinger, and Ronald J. Stouffer. 2009. "Model Projections of Rapid Sea-Level Rise on the Northeast Coast of the United States," *Nature Geoscience*, Vol. 2, pp. 262-266.

[13]Jonathan L. Bamber, Riccardo E.M. Riva, Bert L.A. Vermeersen, and Anne M. LeBrocq. 2009. "Reassessment of the Potential Sea-Level Rise from a Collapse of the West Antarctic Ice Sheet," *Science*, Vol. 324, pp. 901-903.

[14]David T. Pugh. 1996. *Tides, Surges, and Mean Sea-Level,* John Wiley & Sons, Ltd., Chichester, United Kingdom, p. xiv and p. 472.

[15]Roger Flather, Trevor Baker, Phil Woodworth, Ian Vassie, and David Blackman. 2001. "Integrated Effects of Climate Change on Coastal Extreme Events," Proudman Oceanographic Laboratory Internal Document No. 140, p. 20.

[16]S. Nerem, personal communication, April 13, 2010.

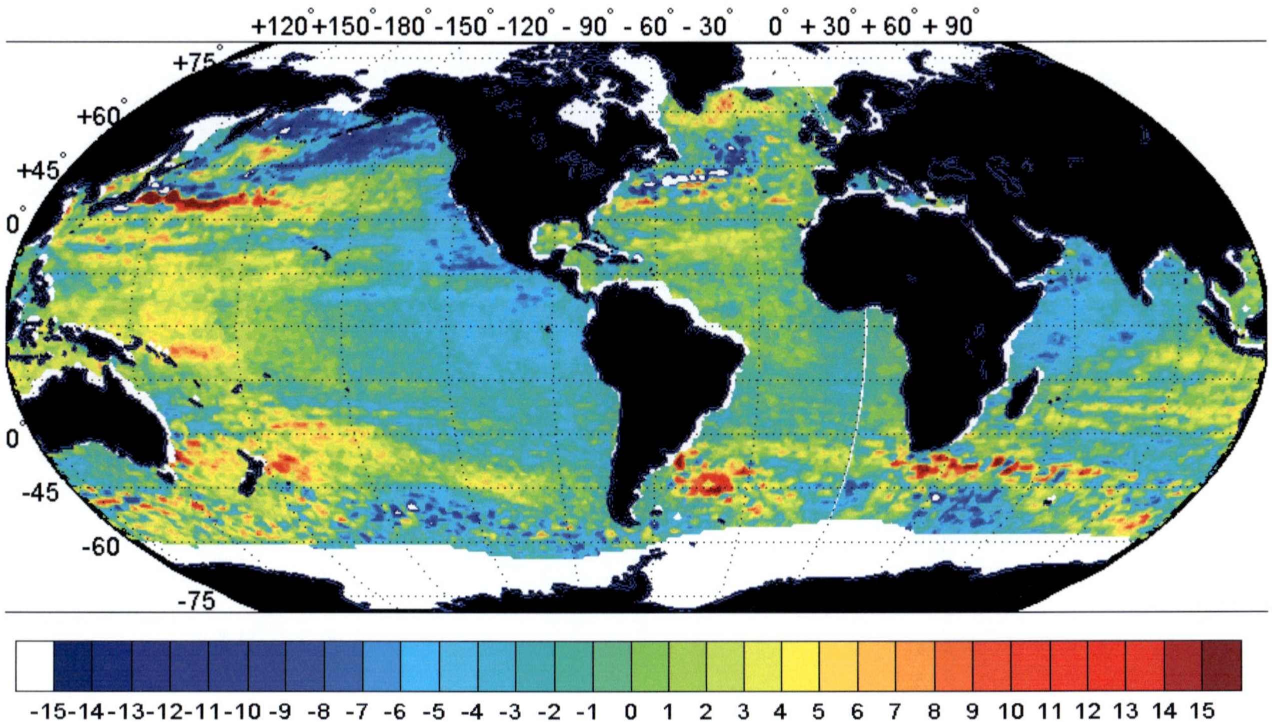

FIGURE 3.2 Observations from the TOPEX-Poseidon altimeter satellite of the global rise in sea level over about 12 years (in mm/yr). SOURCE: Courtesy of NASA/JPL-Caltech. Available at http://www.jpl.nasa.gov/imagepolicy/.

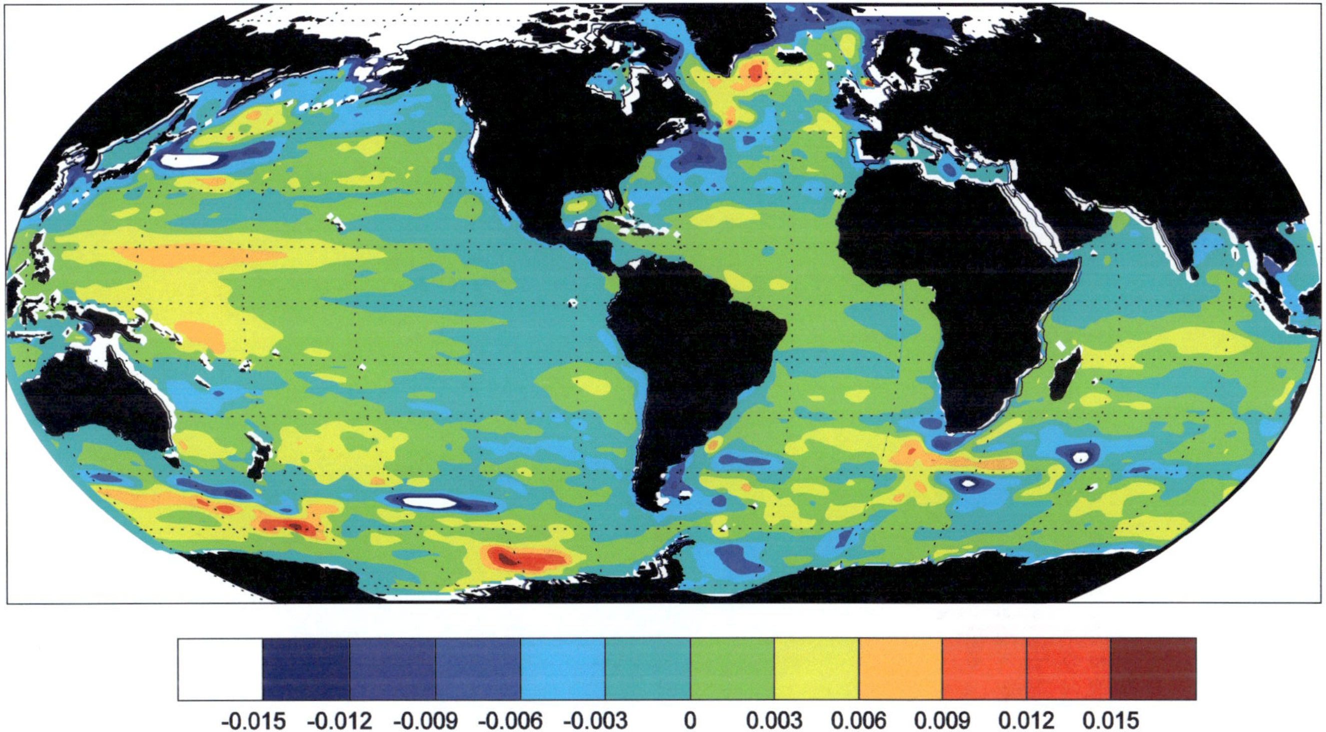

FIGURE 3.3 Estimated sea-level change over 12 years (in m/yr) from a combination of altimetric, in situ observations and a general circulation model (updated from C. Wunsch, R. Ponte, and P. Heimbach, 2007, "Decadal Trends in Sea-Level Patterns: 1993-2004," *Journal of Climate* [pdf]). SOURCE: Courtesy of NASA/JPL-Caltech. Available at http://www.jpl.nasa.gov/imagepolicy/.

mates of a great variety of types. That the two pictures differ in some significant aspects is a reflection of the oceanographic complexity of the problem. The large magnitudes of the regional shifts, much larger than the global mean, lead to the inference that shifts in the wind fields driving the ocean circulation are a major element and would have to be part of any predictive system.

FINDING 3.1: Peer-reviewed literature since the 2007 Intergovernmental Panel on Climate Change *Fourth Assessment Report* (AR4) suggests that loss of ice from small ice bodies (e.g., mountain glaciers and small ice caps) may have been underestimated in the last IPCC report and that major changes in Greenland and Antarctic ice sheet dynamics can take place over relatively short timescales. Sea-level variations caused by shifts in wind, rain, evaporation, and land-ice volume can cause far greater local changes in sea-level variations than the global mean rise that is projected from thermal expansion of the ocean and land-surface melt-water runoff.

RECOMMENDATION 3.1: Based on recent peer-reviewed scientific literature, the Department of the Navy should expect roughly 0.4 to 2 meters global average sea-level rise by 2100, with a most likely value of about 0.8 meter. Projections of local sea-level rise could be much larger and should be taken into account for naval planning purposes. However, U.S. naval leadership (e.g., the Oceanographer of the Navy) should be aware that this estimate is subject to change, and it should be reviewed routinely for any significant change.

Assessing Exposure to Sea-Level Rise

Exposure to damage from future sea-level rise for naval installations will vary from locality to locality. No nation with coastal exposure will be spared the impacts of future sea-level rise; indeed, the viability of some island nations will be at risk if upper ranges of projected sea level materialize over the next few decades. In the aggregate, the effects of sea-level rise and more intense storms on infrastructure and facilities that constitute our built environment will be among the most unequivocal impacts of global climate change.

Each coastal naval installation needs to be examined individually. Thus predictions of greater-than-average sea-level rise along the coast of the Carolinas, coupled with potentially strengthened tropical and subtropical disturbances, would render bases there much more vulnerable than in areas where mean sea level has been falling, such as Diego Garcia. But even the latter would be subject to shifts in tropical cyclone intensity and fluctuations. Generalizations are difficult to make beyond the need to regard each facility as requiring specific scrutiny and assessment of risk.

Increased exposure for coastal regions due to both sea-level rise and intensified coastal storms provides a good example of the value of a well-developed vulnerability analysis and decision making based on risk management principles.

FINDING 3.2: Neither regional nor global sea level is of primary interest in determining naval coastal installation vulnerability. Rather, it is the increased vulnerability associated with extreme events (storm surges) and their dependence on changes in regional sea level, tidal amplitudes, and the nature of extraordinary meteorological forces that are of greatest importance.

RECOMMENDATION 3.2: In performing vulnerability analysis, naval facility managers should recognize that each and every naval facility has a unique configuration and requires ongoing oversight of the changing risks as the climate system shifts. For example, local storm surge impact in climate-induced extreme storm events is likely to represent a bigger vulnerability risk than will sea-level rise alone.

NAVAL COASTAL INFRASTRUCTURE ISSUES

As stated above, global sea-level rise has significant potential to affect many naval coastal installations.[17] These installations are enduring facilities, predominately in the coastal zone, that have been built to last for decades. All were constructed before climate change was recognized as a factor in their design and construction.[18]

The committee reviewed two preliminary assessments of U.S. military coastal installations at risk from coastal inundations caused by sea-level rise.[19]

[17]In this chapter, the definition of infrastructure is limited to U.S. Navy, U.S. Marine Corps, and U.S. Coast Guard buildings, roads and highways, constructed facilities such as piers, docks, and runways, and their supporting utilities. See also Thomas R. Karl, Jerry M. Melillo, and Thomas C. Peterson, 2009, *Global Climate Change Impacts in the United States,* Cambridge University Press, New York, pp. 25-47.

[18]The potential impact of climate change is now being formally recognized in the design and construction of some military civil projects. For example, the U.S. Army Corps of Engineers (USACE) has issued guidance for the incorporation of the direct and indirect physical effects of projected future sea-level change in managing, planning, designing, constructing, and maintaining USACE projects and the system of projects. See Department of the Army, Circular No. 1165-2-211, Washington, D.C., July 2009.

[19]In these studies, the designation of naval military coastal installations as low-, moderate-, high-, or very-high-risk due to sea-level rise is based on the U.S. Geological Survey's (USGS's) Coastal Vulnerability Index (CVI) methodology and ranking techniques. The USGS methodology estimates the risk of weather-related coastal impacts using a set of factors such as the tidal range, wave height, coastal geomorphology, and the historical rate of relative sea-level rise. Risks are assigned as one of four CVI levels (low, moderate, high, and very high) based on histogram plots and visual inspection of the cumulative collected data. For example, in a USGS CVI plot of coastal vulnerability for a targeted area of the U.S. Atlantic Coast, CVI values from 8.7 to 15.6 are considered moderate risk. High-risk values lie between 15.6 to 20.0, and CVI values above 20.0 are classified as very high risk. Source:

Many of the U.S. military installations identified at "very high risk" or at "high risk" are naval installations.[20] These assessments provide a starting point for more in-depth evaluations. As directed by requirements for the Department of Defense's (DOD's) Quadrennial Defense Review (QDR), a broader and more detailed assessment will provide a foundation, but there is a clear need for a more detailed global analysis and an action plan to address the vulnerabilities of those coastal installations identified as being at "very high risk" and at "high risk."[21] The assumptions, decisions, and time lines for addressing these risks should be determined on a consistent basis across the DOD and all naval services. The committee suggests that additional risk factors beyond current indicators of sea-level rise, tidal range, and coastal geomorphology be included in future analyses, including factors such as regional extreme weather history and potential impacts on critical infrastructure—such as communications, transportation, and utilities. In addition, groundwater drawdown and replenishment, saltwater intrusion, and recharge of the aquifer to prevent subsidence are critical factors to take into account in coastal vulnerability analyses. The committee believes that these analyses must explicitly address the potential impact that the combined issues of sea-level rise and more intense storm surges could have on critical military missions.

Some specific impacts that can be anticipated include effects on piers, utilities, and freshwater. Piers will be affected by the sea level relative to the height of the pier as the force exerted by the ship and waves on the pier may grow to be outside the original structural design criteria.[22] This may require significant modifications of piers and waterfront structures at naval shore stations. Utilities such as electrical substations, sewage treatment facilities, and communications nodes, as well as other important and critical infrastructures on the base, may be destroyed or seriously degraded by flooding caused by sea-level rise. Adverse weather can have a deleterious effect on the aboveground utilities system due to high winds,

---

Maj Gen Richard Engel, USAF (Ret.), Director, Climate Change and State Stability Program, National Intelligence Council, "National Intelligence Assessment on the National Security Implications of Global Climate Change to 2030," presentation to the committee, October 19, 2009, Washington, D.C. For additional information on the CVI, the reader is referred to the USGS's National Assessment of Coastal Vulnerability website; available at http://woodshole.er.usgs.gov/project-pages/cvi/.

[20]As one example of the risk, the QDR vulnerability assessment identified 128 DOD installations that could be affected by a sea-level rise of equal to or greater than 1 meter. Fifty-six of these installations (or 43 percent of the total) were Navy installations. This number represents more than 50 percent of the 103 Navy installations that reported. Roughly $100 billion is the estimated dollar value of U.S. Navy installations that are at risk due to this one facet of climate change.

[21]The 2010 Quadrennial Defense Review (QDR) explicitly asked the military services to include climate change trends in its address of the national strategic and security environment. See U.S. Department of Defense, "2010 QDR Terms of Reference Fact Sheet," April 27, 2009, Washington, D.C. The committee reviewed a preliminary report of the coastal infrastructure vulnerability assessment that fed into the QDR.

[22]For a discussion of engineering considerations for coastal structures potentially impacted by sea-level rise, including piers and wharves, see National Research Council, 1987, *Responding to Changes in Sea Level: Engineering Implications,* National Academy Press, Washington, D.C., pp. 106-107.

lightning, and so on. Utilities can be hardened against this type of condition, but at a cost. Saltwater intrusion into aquifers caused by rising sea level where bases draw their freshwater will impact the availability and cost of freshwater.

In addition to facilities management challenges, there are other considerations related to climate change impacts on Navy and Marine Corps installations. Missions of various naval installations may be affected by the increased requirement to support contingency operations in the humanitarian assistance/disaster relief (HA/DR) area. The impact will be felt mostly in the area of logistics. Although bases have contingency plans for many kinds of events, the plan for evacuation due to flooding may require revision as well as more frequent assessment than is required now. Communities surrounding naval installations may also be stressed and require contingency plans, including the need to address potential impacts to coastal wetlands and ecosystems and on local public health.[23,24]

CURRENT NAVAL COASTAL VULNERABILITY STUDIES

There are currently at least three separate Navy groups involved in the analysis of coastal installation vulnerability issues for the Navy: the Naval Space Warfare Systems Command (SPAWAR), the Naval Facilities Engineering Service Center (NAVFESC),[25] and the Naval Installations Command.[26] Additionally, prior to the 2010 QDR the U.S. Marine Corps began conducting an analysis of

[23]For example, rapid sea-level rise and intense storm surges can cause segmentation of barrier islands and disintegration of wetlands, each having societal consequences. A comprehensive review of how sea-level rise can affect coastal environments is provided in the 2009 U.S. Climate Change Science Program (CCSP) Synthesis and Assessment Product 4.1 (SAP 4.1), *Coastal Sensitivity to Sea-Level Rise: A Focus on the Mid-Atlantic Region,* U.S. Climate Change Science Program, Washington, D.C., January.

[24]A case in point was the response of the health sector along the Gulf Coast of the United States after the 2005 Hurricane Katrina. As noted in a 2007 National Research Council report on public health impact of disasters, the aftermath of Hurricane Katrina consisted of short-, medium-, and long-term responses specific to saving lives, controlling health hazards, and reconstruction efforts, respectively, across federal, state, and local agencies. In terms of reducing vulnerability for drinking water, vector diseases in flooded areas, and indoor air quality, and so on, it was noted that a risk assessment needs to be performed specific to the region that considers enhanced monitoring, precautions to reduce risk, and effective communication strategies. See National Research Council, 2007, *Environmental Public Health Impacts of Disasters: Hurricane Katrina, Workshop Summary,* Roundtable on Environmental Health Sciences, Research and Medicine, The National Academies Press, Washington, D.C.

[25]See, for example, Kathleen Paulson and Dallas Meggitt, 2008, *US Naval Facilities Engineering Service Center Environmental Program on Climate Change,* Naval Facilities Engineering Service Center, Port Hueneme, Calif.

[26]CAPT Brant Pickrell, CEC, USN, Capital Investments Branch Head, Shore Readiness Division (N46), Deputy Chief of Naval Operations for Fleet Readiness and Logistics (OPNAV N4), "Preliminary Climate Change Related Naval Base Assessments—A Status Report," presentation to the committee, October 19, 2009, Washington, D.C.

its coastal vulnerabilities.[27] Also, the DOD's Strategic Environmental Research and Development Program (SERDP) has initiated climate change military infrastructure studies.[28] This committee believes that to avoid duplication of effort and to ensure a more comprehensive and consistent assessment, a more coordinated vulnerability analysis is needed across the naval installations nationally and internationally. This effort should also be coordinated with vulnerability assessments for other military installation across the DOD.

The Office of the Chief of Naval Operations, N46 (Shore Readiness), informed the committee that the Navy is reviewing the Military Construction Program to evaluate the impact of climate change on the facilities in the program. Life-cycle costing and impacts of climate change, energy reduction, and reduction of greenhouse gases can all be considered in investment decisions by the services. This approach is also recommended for non-DOD naval forces in their capital investment decisions.

Considering the current measurements for sea-level rise, it is not anticipated that the Navy will need to make a major resource investment in the near term, with the exception of those naval installations currently identified as being at very high risk. However, on the longer time horizon (the next 10 to 20 years), investments will have to be made for adaptation and mitigation of climate impacts at many naval coastal installations, and those investments may have implications for decisions being made today.[29]

As the mission of naval forces changes to meet the demands of climate change, it is also suggested that the Navy team review the potential requirement for new bases. This would include new enduring bases in places like the Arctic, the increased ability to logistically support sea basing of efforts much like the recent efforts in Haiti during earthquake recovery, and, finally, new requirements for contingency bases in response to world situations that may take the Navy to new places like Iraq and Afghanistan. Even though all these types of basing are within the current capability of the U.S. Navy, the impact of climate change should be routinely evaluated for all future naval base decisions.

A 1987 National Research Council report explores engineering considerations for dealing with sea-level rise and recommends a multiple-scenario

---

[27]Elmer W. Ransom, Environmental Management Section, Headquarters, U.S. Marine Corps, and Capt Anthony V. Ermovic, USMC Facilities Branch Head, Headquarters, U.S. Marine Corps, "Marine Corps Perspectives and Climate Change Initiatives," presentation to the committee, September 18, 2009, Washington, D.C.

[28]The Office of the Secretary of Defense's Strategic Environmental Research and Development Program currently sponsors several projects related to the assessment of the impact of global sea-level rise on military infrastructure. These projects are managed under SERDP's Sustainable Infrastructure Projects Program. Descriptive information on these projects (SI-1700, -1701, -1702, and -1703) is available at http://www.serdp.org/Research/SI-Facilities-Management.cfm. Accessed April 17, 2010.

[29]The Navy's investment needs, if any, would be reflected in Program Objective Memorandum (POM) submissions. The POM submission is a 5-year outlook on budget requirements. It starts with the year following the President's Budget, which is always 1 year ahead of the current year.

approach to deal with uncertainties for which no reliable or credible probabilities can be obtained. While this 1987 report is dated from the observed sea-level rise data that it presents, its recommended approaches to deal with sea-level changes remain valid and should be considered by naval facilities management.[30]

FINDING 3.3: U.S. Navy, Coast Guard, and Marine Corps coastal installations around the globe will become increasingly susceptible to projected climate change. Several assessments now under way on naval installation vulnerabilities appear to be focused primarily on static sea-level rise and coastal inundation only. According to these current assessments, some adaptive actions are indicated owing to already identified vulnerabilities at specific naval installations. The preliminary review of climate-change-related base vulnerabilities across the DOD—currently under way as directed by the 2010 Quadrennial Defense Review[31]—does not include some important factors that affect coastal installation vulnerabilities, although it provides a baseline assessment across all branches of the armed services and serves as a starting point for more in-depth analysis and action.

RECOMMENDATION 3.3a: The Commander, Naval Installations Command, and the Navy Director for Fleet Readiness and Logistics should work with their U.S. Coast Guard and Marine Corps counterparts—and in conjunction with the other armed services and the Office of the Secretary of Defense—to ensure that a coordinated analysis is undertaken to address naval-installation vulnerability to rising sea levels, higher storm surges, and other consequences of climate change. In performing this vulnerability analysis, naval facility managers should recognize that each and every naval facility has a unique configuration and requires ongoing oversight of the changing risks as the climate system shifts. For example, local storm surge impact in climate-induced extreme storm events is likely to represent a bigger vulnerability than sea-level rise alone.

RECOMMENDATION 3.3b: For Program Objective Memorandum (POM)-14 planning purposes, the Chief of Naval Operations should prepare to invest in early-stage adaptation for targeted low-elevation naval installations identified in current vulnerability assessments as being at "very high risk" from more intense storm surges, sea-level rise, and other climate change impacts. Other risks for naval installations as a result of projected climate change require further analysis and planning at this time, but no immediate direct additional substantial investment beyond current budget plans.

---

[30]The National Research Council. 1987. *Responding to Changes in Sea Level: Engineering Implications,* National Academy Press, Washington, D.C.

[31]Secretary of Defense (Robert M. Gates). 2010. *Quadrennial Defense Review,* Department of Defense, Washington, D.C., February.

FINDING 3.4: The U.S. military is well aware of the risks to its coastal facilities and infrastructure from sea-level rise. Recent observations of sea-level rise have exceeded projections made only a decade earlier, and the increasing realization of the potential of changes in ice dynamics leads to the further realization that there perhaps continues to be underestimation of the sea-level rise that would be associated with likely future climate change. The risk of harm to military and civilian coastal facilities from sea-level rise is not linear with the rate of rise. There will be thresholds at which existing natural and built coastal barriers are exceeded. An important dimension of this risk is that of storm surge, especially if warmer future conditions give rise to an increased intensity of storms.

RECOMMENDATION 3.4: For risk management purposes, U.S. naval leaders would be prudent to err on the side of overestimation of future sea-level rise when renovating existing or planning new coastal facilities. The Navy and other branches of U.S. services that have historic commitments to HA/DR efforts for the United States and beyond need to consider as highly probable the need to enhance these capabilities to be prepared for increased damage from coastal storms.

4

Allied Forces' and Partners' Issues

U.S. allies and their militaries will face climate-change-related issues similar to the challenges that the United States and its naval forces will face. Demands are expected to increase for humanitarian assistance/disaster relief (HA/DR) and maritime security missions. In some cases, potential Arctic engagement may be necessary, as climate change influences the geopolitical landscape. However, internal economic and political pressure, as well as geographic proximity to climate-change-influenced geopolitical "hot spots," will lead to different responses from these allies and partners. Some allies will have an inherently greater capacity than others, and some may be required to deal with severe local climate-change-related issues internally or just across their borders. This chapter will examine these issues from the perspective of potential strategies for U.S. naval forces to form partnerships and develop cooperative approaches in planning for global climate-change-related issues beyond the U.S. borders.

This chapter begins with an overview of global climate change effects that have the potential to require U.S. naval responses. The chapter then focuses on how these geographic hot spots may affect U.S. allies, partners, and other nations, and it examines recent HA/DR efforts in Haiti as an illustrative case study. The chapter concludes with a discussion of regional vulnerabilities and specific findings and recommendations toward developing maritime partnerships as central to cooperative strategies for climate-change-related adaptation and planning, including suggested partnerships in the Arctic region.

The World Bank's *World Development Report 2010* states that all regions of the world are vulnerable to climate change.[1] Some have more natural susceptibility

[1]The World Bank. 2009. *World Development Report 2010: Development and Climate Change*, November, The World Bank, Washington, D.C.

to climate effects, however, and many have a lower capability to adapt. Possible effects in these areas include drought, flood, mass migrations, conflict, and humanitarian disasters. The confluence of these factors will most likely present challenges for the United States and its allies. According to the National Intelligence Council (NIC), migrants fleeing natural disasters in North Africa, for example, may move in large numbers into NATO countries in southern Europe.[2] Such mass migrations are likely to challenge the physical and social infrastructure in countries of origin and in recipient countries. While migration may or may not be seen as a security challenge, contending with such events is likely to place demands on the military and maritime resources of partner nations, as it has at times in the past.[3]

Taking natural and human-made vulnerability into account, the committee found that there were several global hot spots of particular concern to the United States and its allies. The "hot spot" concept has been cited by both the World Bank, in its development report, and the NIC and is expanded upon in this chapter.

IMMEDIATE CHALLENGES ASSESSMENT: ALLIED FORCES' CLIMATE-CHANGE-RELATED ISSUES

Given the judgment that climate change will result in a range of effects for all nations, U.S. military forces, particularly naval forces, are likely to contend with climate-related contingencies around the world, as described in Chapter 2. This is both a reflection of U.S. global economic and security interests and the fact that U.S. maritime forces are forward-deployed around the world and likely to be "first responders" in contingencies requiring a U.S. response. The pervasive nature of these challenges has important implications for U.S. relations with allied and partner maritime forces.

First, climate change will affect U.S. allies in varying ways domestically and regionally. While these challenges are unlikely to trigger any treaty obligations (under NATO, ANZUS [the Australia, New Zealand, United States Security Treaty], or the U.S.-Japan Security Treaty, for example), it is very likely that allies may request U.S. assistance, particularly in dealing with humanitarian assistance, disaster relief, and mass migration. Traditionally, the posture of the United States has been to assist allies to the greatest extent possible.

Second, given the historical record of U.S. military support for global humanitarian and disaster relief operations, the President of the United States is likely to continue directing U.S. maritime forces to respond to climate change contingencies in hot spots around the globe. The capabilities and willingness of allies and partners to participate in these responses will be critical because the

---

[2]See National Intelligence Council, 2008, *2025 Global Trends Report,* November, p. 53; available at http://www.dni.gov/nic/PDF_2025/2025_Global _Trends_Final_Report.pdf. Accessed May 25, 2010.

[3]Examples include Operation Sea Signal in 1994 and Operations Safe Harbor and Able Manner in 1991-1992.

committee judges that the United States will lack the resources and, in some instances, the strategic justification for responding alone to every request for assistance in dealing with climate-related contingencies, even when U.S. interests may be directly at stake.

The Haiti Earthquake Response

The response to the January 2010 earthquake disaster in Haiti provides some insights into the role that U.S. naval forces may be expected to play in future international HA/DR climate-related contingencies. Although the earthquake was not a climate-related event, there would very likely be operational similarities to climate-related disasters; therefore, this incident may be instructive for future naval missions. A hallmark of the January 2010 operation was the U.S. Navy's cooperation not only with other U.S. military services but also with U.S. allies, the United Nations, nations with no formal military ties to the United States, and private organizations. A preliminary report of the lessons learned in Haiti includes the following concerns:[4]

- Balance the Push Versus Pull of Forces:[5] *Quick initial deployment is critical. However, once local needs are determined, better coordination is needed to assure the proper balance between pushing troops and solutions onto local commands, versus the pull of forces as needed.*
- Coordination with Non-Governmental Organizations (NGOs) Is Critical: *NGOs are critical partners in HA/DR operations, providing relief along with local government resources once naval personnel missions are complete. U.S. naval personnel should continue to build on its relationships and formal programs with NGOs.*
- Preplan for Strategic Communications Needs: *Due to the interagency and international scope of the effort, strategic communications and coordinated post-mission withdrawal plans are needed, including preplanning and coordination with the Department of State.*
- Improve Inbound Cargo Coordination: *To help avoid misrouting and improve efficiency, formal coordination with stakeholders should be established for handling of inbound cargo, including any special handling requirements.*
- Improve Medical Planning/Coordination: *Early arrival of experienced medical personnel and medical planners is critical. Navy hospital ships are indispensable, but depending on location of the crisis, their arrival may take weeks.*

---

[4]CAPT Alfred Collins, USN, Chief of Staff, Fourth Fleet, U.S. Navy Southern Command, "Haiti HA/DR and Climate Change Impact on Naval Operations in SOUTHCOM AOR," presentation to the committee, March 23, 2010.

[5]A push-pull system in logistical supply situations describes the movement of a product (in this case, personnel) between two subjects. The consumers (i.e., local commands) usually *"pull"* the products they demand for their needs, while the suppliers (i.e., command headquarters) *"push"* them toward the consumers.

The formal Department of Defense (DOD) lessons-learned report from Haiti is anticipated to elaborate on these and other items, and it can serve as a basis for future preplanning of international HA/DR activities between the United States, its allies, and other partners.

The Arctic

In addition to the HA/DR issues for the United States and its allies, the opening of the Arctic has the potential to be a new "great game" in geostrategic terms and thus serves as a challenge for U.S. and NATO forces.[6] The potential challenges for alliances and other bilateral and multilateral relationships range from competition for Arctic resources, to navigation rights through the area, to which nation has responsibility and capability for search and rescue in the region. At the most extreme, conflicts or tensions over sovereign rights and jurisdiction in the Arctic may remain sensitive issues over the next 20 years. In addition to shifting the relationships of "frontline" Arctic nations (Canada, Denmark, Finland, Iceland, Norway, Russia, Sweden, and the United States), the opening of the area will affect global shipping routes, which in turn may affect U.S. bilateral and multilateral strategic and economic relationships around the world, with implications for maritime forces.

KEY GEOGRAPHIC "HOT SPOT" PROJECTIONS, MIGRATION PATTERNS, AND CLIMATE CHANGE IMPACT ASSUMPTIONS

National Intelligence Council Assessments

In follow-up analysis to its 2008 report *National Security Implications of Global Climate Change to 2030*,[7] the NIC embarked on a research effort to explore in greater detail the national security implications of climate change in six countries/regions of the world: (1) Russia; (2) China; (3) Southeast Asia and the Pacific Islands states; (4) India; (5) Mexico, the Caribbean, and Central America; and (6) North Africa. The committee has reviewed the analysis provided

---

[6]The "great game" is a term originally used to describe the strategic rivalry between the British Empire and the Russian Empire to control major portions of Eurasia in the 19th century. Some political historians have suggested that a contemporary version of the great game international rivalry has been played out in the Middle East and the Balkans since the fall of the former USSR and the end of the Cold War. The great game terminology has also been used by some writers and observers of the Arctic. For example, see *Great Game in a Cold Climate: Canada's Arctic Sovereignty in Question,* Canada National Defence website; available at http://www.journal.forces.gc.ca/vo6/no4/north-nord-01-eng.asp. Accessed June 4, 2010.

[7]See House Permanent Select Committee on Intelligence and House Select Committee on Energy Independence and Global Warming, "Statement for the Record by Dr. Thomas Fingar, Deputy Director of National Intelligence for Analysis—National Intelligence Assessment on the National Security Implications of Global Climate Change to 2030," June 25, 2008.

in these reports.[8] As of May 2010, only the first of three planned phases of data is available; the first phase assesses merely the physical impacts of climate change on these key countries, not the socioeconomic impacts or the national security impacts. The Phase I data are summarized in Table 4.1. In essence, the story that the Phase I data tell is that most areas of the world are likely to experience water stress (including floods) and a range of effects on coastal areas, with the potential for serious secondary effects (such as effects on availability of energy or agricultural productivity).

The countries and regions examined in Phase I are of particular strategic concern to the United States. According to the NIC, India and China are especially vulnerable to climate change, particularly given the size of their populations and existing development challenges. An important finding is that although Russian authorities may believe that Russia will have net gains from a warming climate (by gaining access to Arctic resources, for example), there is evidence that Russia will contend with serious challenges, particularly to its energy sector, as permafrost thaws earlier and deeper—impeding construction of new production areas. This could have a material negative impact on Russia's oil and gas industry, the single greatest source of income to the Russian state. The Americas and North Africa are likely to see conditions that will continue or increase current migration patterns. The remaining phases of the NIC's climate change work will assess state instability issues within the targeted region and security implications for the United States, including work to provide a more quantitative assessment (see Box 4.1).

World Bank Regional Climate Change Vulnerability Assessments

In work similar to the NIC Phase I assessments, the World Bank *World Development Report 2010* presents a projection of global climate-change-related vulnerabilities.[9] The report suggests specific vulnerabilities in six global regions that may be of importance for U.S. forces or their allies: (1) Sub-Saharan Africa, (2) East Asia and Pacific, (3) Europe and Central Asia, (4) Latin America and the Caribbean, (5) Middle East and North Africa, and (6) South Asia. These regional climate change vulnerabilities from the World Bank report are summarized below:

1. *Sub-Saharan Africa.* Sub-Saharan Africa is reported to suffer from natural fragility (two-thirds of its surface area is desert or dry land) and has high exposure to droughts and floods, which are forecast to increase with further climate change. The region's economies are highly dependent on natural resources. Rain-

[8]National Intelligence Council. 2008. *The Impact of Climate Change to 2030,* a series of commissioned research reports and conference reports. See http://www.dni.gov/nic/special_climate2030.html. Accessed April 8, 2011.

[9]The World Bank. 2009. *World Development Report 2010: Development and Climate Change,* November, The World Bank, Washington, D.C.

TABLE 4.1 Summary of National Intelligence Council Projected High-Risk Impacts of Climate Change to the Year 2030, by Country or Region

	High-Risk Impacts					
Country or Region	Coastal Regions	Water Resources	Agriculture	Energy	Migration	Socioeconomic/ Political Stress
Russia		X	X	X	X	X
China	X	X				X
Southeast Asia and Pacific Islands	X	X	X			
India	X	X	X	X	X	X
Mexico, the Caribbean, and Central America	X	X	X	X	X	
North Africa		X	X		X	X

SOURCE: Based on data in National Intelligence Council, 2008, *The Impact of Climate Change to 2030,* a series of commissioned research reports and conference reports. See http://www.dni.gov/nic/special_climate2030.html. Accessed April 8, 2011.

fed agriculture contributes some 30 percent of gross domestic product (GDP) and employs about 70 percent of the population. Inadequate infrastructure could hamper adaptation efforts, with limited water storage despite abundant resources. Malaria, already the biggest killer in the region, is spreading to higher, previously safe, altitudes.

2. *East Asia and Pacific.* In East Asia and the Pacific, one major driver of climate change vulnerability is the large number of people living along the coast and on low-lying islands: more than 130 million people in China, and roughly 40 million, or more than half the entire population, in Vietnam. A second driver is the continued reliance, particularly among the poorer countries, on agriculture. As pressures on land, water, and forest resources increase—as a result of population growth, urbanization, and environmental degradation caused by rapid industrialization—greater variability and extremes will complicate their management. In the Mekong River Basin, for example, the rainy season will see more intense precipitation, while the dry season will lengthen by 2 months.

3. *Europe and Central Asia.* Vulnerability to climate change in Eastern Europe and Central Asia is driven by a lingering Soviet legacy of environmental mismanagement and the poor state of much of the region's infrastructure. As an example: rising temperatures and reduced precipitation in Central Asia will exacerbate the already negative impact of the disappearing Southern Aral Sea (caused by the diversion of water to grow cotton in a desert climate), while sand and salt from the dried-up seabed are blowing onto Central Asia's glaciers, accelerating the melting caused by higher temperatures. Poorly constructed, badly maintained, and aging infrastructure and housing are ill suited to withstand storms, heat waves, or floods.

4. *Latin America and the Caribbean.* Latin America and the Caribbean's most critical ecosystems are under threat. First, the tropical glaciers of the Andes are expected to disappear, changing the timing and intensity of water available to several countries; this, in turn, will result in water stress for at least 77 million people as early as 2020 and will threaten hydropower, the source of more than half the electricity in many South American countries. Second, warming and acidifying oceans will result in more frequent bleaching and possible diebacks of coral reefs in the Caribbean, which host nurseries for an estimated 65 percent of all fish species in the basin, provide natural protection against storm surge, and are a critical tourism asset. Third, damage to the Gulf of Mexico's wetlands will make the coast more vulnerable to more intense and more frequent hurricanes. Fourth, the most disastrous impact could be a dramatic dieback of the Amazon rain forest and a conversion of large areas to savannah, with severe consequences for the region's climate.

5. *Middle East and North Africa.* Water is the major vulnerability in the Middle East and North Africa, the world's driest region, where per capita water availability is predicted to halve by 2050 even without the effects of climate change. The region has few attractive options for increasing water storage, since

BOX 4.1
National Security and the Ranking of Global Climate Change Adaptive Capacity

Researchers have recently taken on the challenge of assessing adaptive capacity in a comparative quantitative framework. In this work, a comparative study of country-specific resilience to climate change is provided based on the Vulnerability-Resilience Indicators Model (VRIM).[a] A representative preliminary VRIM comparison of a group of 10 countries (from a 160-country database) is indicated in Figure 4.1.1 for base year 2006. Additional detailed views of key components of adaptability are also available from the model.

[a]VRIM is a hierarchical model with four levels and eight sectors. Each of the hierarchical-level values is composed of the geometric means of participating values. Proxy values are indexed by determining their location within the range of proxy values over all countries or states. The final calculation of resilience is the geometric mean of all eight sectors. The vulnerability index (level 1) is derived from two indicators (level 2): sensitivity (how systems could be damaged by climate change) and adaptive capacity (the capability of a society to maintain, minimize loss of, or maximize gains in welfare). Sensitivity and adaptive capacity, in turn, are composed of sectors (level 3). For adaptive capacity, these sectors are human resources, economic capacity, and environmental capacity. For sensitivity, the sectors are settlement/infrastructure, food security, ecosystems, human health, and water resources. Each of these sectors is composed of one to three proxies (level 4). The proxies under adaptive capacity are as follows: human resource proxies are the dependency ratio and literacy rate; economic capacity proxies are gross domestic product (GDP) (market) per capita and income equity; and environmental capacity proxies are population density, sulfur dioxide divided by state area, and percent of unmanaged land. Proxies in the sensitivity sectors are water availability, fertilizer use per agricultural land area, percent of managed land, life expectancy, birthrate, protein demand, cereal production per agricultural land area, sanitation access, access to safe drinking water, and population at risk due to sea-level rise.

close to 90 percent of its freshwater resources are already stored in reservoirs. The increased water scarcity, combined with greater variability, will threaten agriculture, which accounts for some 85 percent of the region's water use. Vulnerability is compounded by a heavy concentration of population and economic activity in flood-prone coastal zones and by social and political tensions that resource scarcity could heighten.

6. *South Asia.* South Asia suffers from an already stressed and largely degraded natural resource base resulting from geography coupled with high lev-

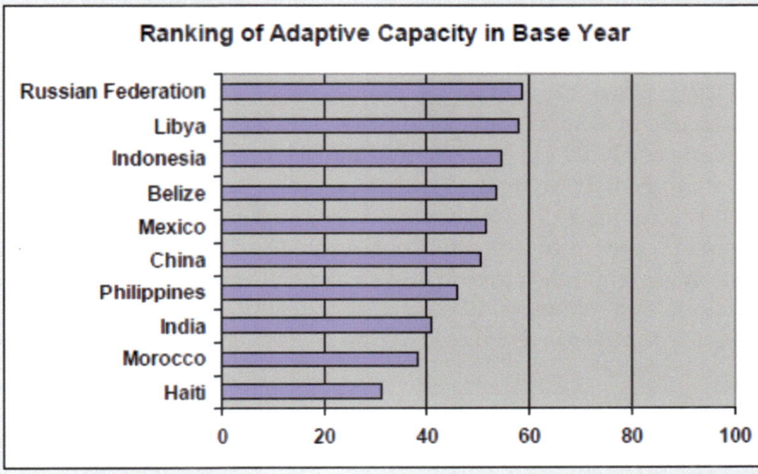

FIGURE 4.1.1 Results from VRIM models have been used in regional vulnerability analysis conducted by the National Intelligence Council. Additional details on VRIM and its application to adaptability studies in six regions of the world are found at http://www.dni.gov/nic/special_climate2030.html, accessed April 8, 2011. SOURCE: Elizabeth L. Malone, Joint Global Change Research Institute, "Scientific Knowledge About Climate Change for Consideration in National Security Planning," presentation to the committee, February 4, 2010.

els of poverty and population density. Water resources are likely to be affected by climate change through its effect on the monsoon, which provides 70 percent of annual precipitation in a 4-month period, and on the melting of Himalayan glaciers. Rising sea levels are a dire concern in the region, which has long and densely populated coastlines, agricultural plains threatened by saltwater intrusion, and many low-lying islands. In more severe climate change scenarios, rising seas would submerge much of the Maldives and inundate 18 percent of Bangladesh's land.

Based on these analyses, no region of the world is immune to potential climate change impacts, and each region has the potential to generate climate-related missions for U.S. naval forces or U.S. allies and partners. Related to this, the committee also received briefings associated with issues surrounding water availability and conflict.[10] There is growing regional competition for water due to rising populations and rising demands from many sectors around the globe. For example, several African countries are arguing over water rights to the Nile based on claims exerted by Egypt; Israel and Jordan have competing claims to the Jordan River; across the Himalayas, China's dam on the Yarlung Tsangpo River is causing anxieties about water availability in India's northeastern sector and in Bangladesh; and India's own projects to build hydroelectric dams along the Indus River to trap Himalayan waters have caused increased tension with Pakistan.[11] While this committee did not focus on water challenges directly, challenges to water systems and water availability exacerbated by climate change could add to global tensions and lead to potentially broader national security implications and implications for naval forces. Climate-change-related water tensions are a special subset of climate change and should remain on the radar for U.S. national security and naval leaders.[12]

PRELIMINARY STRATEGIES/OPPORTUNITIES TO LEVERAGE U.S. AND ALLIED FORCES AND CAPABILITIES

Given the scope and scale of potential climate change contingencies and vulnerabilities, the United States lacks the resources and capabilities to respond to all plausible scenarios that may directly or indirectly affect the homeland, allies, or general global catastrophic situations. The capabilities and cooperation of partners and allies will not only be important, they will be necessary.

The United States should place a high priority on cooperating with allies, non-allied partners, and private organizations in both anticipating and responding to global climate change and geographic hot spots. The committee agrees that these partnerships at this time are either not sufficiently robust or tailored for the quantity and type of missions that are most likely to occur.

---

[10]Kathy Jacobs, Deputy Director, White House Office of Science and Technology Policy, "Perspectives from OSTP on Water, Adaptation, and the National Assessment," presentation to the committee, February 5, 2010; see also Peter H. Gleick, President, The Pacific Institute, "Water, Climate, and International Security: Definitions, History, and Future Risks," presentation to the committee, November 19, 2009.

[11]Lydia Polgreen and Sabrina Tavernise, "Water Dispute Increases India Pakistan Tension," *New York Times,* July 20, 2010; available at http://www.nytimes.com/2010/07/21/world/asia/21kashmir.html?_r=1&adxnnl=1&ref=water&adxnnlx=1284742861-5IiKNw+K65TaQM6PEasqgQ. Accessed February 14, 2011.

[12]A map of projected change in regional water availability may be found at *World Development Report 2010: Development and Climate Change,* November 2009, The World Bank, Washington, D.C., p. 148.

In a review associated with this study, the committee studied the 2008 National Research Council's Naval Studies Board report entitled *Maritime Security Partnerships.*[13] In the committee's review, it became clear that many of that study's rationales for and proposals concerning maritime security partnerships are pertinent for dealing with future climate-related contingencies, particularly in those requiring HA/DR missions. The committee recommends that the leaders of U.S. naval forces should pay particularly close attention to three of the recommendations (summarized below) from that study when considering climate change:

- Continue bilateral and multilateral training and exercising of U.S. naval personnel with partner nation personnel in maritime security, search and rescue, and HA/DR exercises;
- Explore the expansion of a robust foreign area officer (FAO) program within the Navy to meet the needs of staffing and expanding maritime security partnerships. In addition, the Commandant of the Coast Guard should establish an FAO program and the Commandant of the Marine Corps should expand its present limited FAO program for the development of bilateral and multilateral relationships; and
- Direct the United States Coast Guard to forward deploy Coast Guard cutters to locations that offer opportunities for the joint enforcement of maritime security. These cutters would help to attain Navy and combatant commander engagement goals and would be the correct security assets to employ to meet theater cooperation goals.[14]

With stronger partnerships and more capable partners, the United States will be more likely to mount effective responses to the range of projected climate-related contingencies. Even with better partnerships, however, the United States will not be able to respond to every scenario. Moreover, many partner nations may be unable to commit resources to a catastrophic event because they are fully engaged in their own domestic or regional issues brought on by the same event.

The Department of the Navy, in cooperation with other military services, the Department of Homeland Security, the Department of State, and other relevant agencies, should therefore invest resources in understanding the human impacts of climate change in order to prepare for and prioritize the most plausible contingencies. The DOD's 2010 *Quadrennial Defense Review* directs that the department pursue risk management strategies.[15] In this committee's opinion, it will be very important for the department to apply that recommendation to climate change.

[13]National Research Council. 2008. *Maritime Security Partnerships,* The National Academies Press, Washington, D.C.

[14]Ibid.

[15]Secretary of Defense (Robert M. Gates). 2010. *Quadrennial Defense Review,* Department of Defense, Washington, D.C., February, pp. 84-89.

The DOD should consider climate change to be comparable to other challenges to U.S. interests, focusing planning and strategy on the contingencies that are most threatening for U.S. security.

It is possible that such careful consideration of climate change challenges may result in a determination that elements of the U.S. government other than the naval forces will need to take the lead on climate change response. For example, making U.S. and global communities more resilient to projected changes may be a more appropriate mission for development, aid, or trade agencies rather than military organizations. Engagement and preplanning with leading nongovernmental organizations specializing in HA/DR at the planning table are also highly encouraged.

NATO could become a focal point for leading international military HA/DR efforts, but as of the writing of this report, NATO does not yet have a formal climate change policy.[16] The committee's discussions with senior military officials suggest that many NATO countries have strong national climate change policies, but they lack sufficient capabilities to prepare for or respond to projected climate changes at home and around the world. Although differences of opinion on climate change have at times been divisive in relations among NATO countries, a common effort to develop capabilities and capacity for climate response has the potential to strengthen the alliance.

FINDING 4.1: All regions of the world will experience the effects of projected climate change. Some climate change effects, such as changes in storm patterns and drought, will have direct impacts in the United States. Should regional storms and droughts intensify over time they may well drive mass migrations to the United States from neighboring countries, including Mexico, the Caribbean, and Central America. Projected climate change will also directly and indirectly affect most U.S. allies, including NATO countries, Australia, Japan, and all other major non-NATO allies, which in turn may request or require U.S. assistance.

RECOMMENDATION 4.1: Given that U.S. naval forces cannot be fully prepared for or respond to all plausible climate contingencies, the Chief of Naval Operations, working with the combatant commanders, the Commandant of the Coast Guard, and the Commandant of the Marine Corps, should develop or expand maritime partnerships with other nations. Projected climate change will affect all regions of the world, and so U.S. naval forces should seek to develop these partnerships with long-standing allies and nontraditional partners alike, including Russia, China, and nongovernmental organizations. In particular, developing

[16]See "NATO Secretary General Debates Climate Change Security Threats in Copenhagen," *NATO News,* December 15, 2009; updated April 14, 2010. Available at http://www.nato.int/cps/en/SID-614123F7-2989961A/natolive/news_60163.htm?selectedLocale=en. Accessed June 4, 2010.

climate change response capabilities within the NATO alliance could strengthen global climate change response capabilities and the alliance itself.

THE NEW "GREAT GAME"

The Arctic region covers some 8,100,000 square miles, with volatile weather and very harsh, rapidly changing conditions. Operations in the area, as covered in earlier chapters, are expensive and difficult and require significant and unique resources and training. Changing Arctic conditions are already reshaping geostrategic relationships, including for non-Arctic nations. Indeed, a number of other nations possess Arctic capabilities that exceed those of the United States, and not all of these nations are allies or even frontline Arctic states.

There are eight "frontline" Arctic nations—Canada, Denmark, Finland, Iceland, Norway, Russia, Sweden, and the United States—many with unresolved claims in the region. In addition, Russia, Canada, Norway, and Denmark are all expanding their Arctic military capabilities.[17] Earlier this year, the Russian Security Council posted on its website a paper describing the country's Arctic strategy. The document calls for a new military force to be established by 2020 to protect Russian interests in the region. The Russian strategy also calls for building up military units to secure Arctic coastal borders.[18] Likewise, Canada's "Northern Strategy" documents, published in September 2009, emphasize border protection and the exercise of Canada's sovereignty over its Arctic lands and waters. Norway, Sweden, and Finland have banded together in the Nordic Defense Cooperation Initiative, in part to share and coordinate military resources in the region.[19]

The United States has cooperated routinely with all of these nations on Arctic matters. This has been done on a bilateral basis and through NATO, as well as through the Arctic Council, scientific partnerships, and ad hoc arrangements. Thus far, disagreements on regional issues have been resolved without conflict.[20] Related to this, the committee held discussions on anticipated Arctic issues and

[17]Noel Brinkerhoff. 2009. "U.S. Navy Prepares for Militarization of the Arctic," *All Government*, November 30.

[18]See Katarzyna Zysk, 2010, "Russia's Arctic Strategy, Ambitions and Constraints," *Joint Forces Quarterly*, Issue 57, 2nd Quarter. See also "New Russian Maritime Strategy Highlights Arctic"; available at http://www.barentsobserver.com/new-russian-maritime-strategy-highlights-arctic.4554994-116320.html. Accessed June 4, 2010.

[19]See Canada's Northern Strategy documents at http://www.northernstrategy.gc.ca/index-eng.asp. Accessed February 14, 2011. The committee was also briefed by Ross Graham, Director General, Defence Research and Development Canada, Center for Operational Research and Analysis, February 4, 2010. Norway's Arctic strategy was presented to the committee on March 22, 2010, by MajGen Tom H. Knutsen, Defense Attaché, Royal Norwegian Embassy, Washington, D.C.

[20]In discussions on March 5, 2010, with this committee, ADM James G. Stavridis, USN, Commander of the United States European Command and NATO's Supreme Allied Commander Europe (SACEUR), stated that the United States and NATO, while aware of areas of disagreement with Russia, will seek a cooperative strategy with Russia in the Arctic region.

strategies with government and military representatives of Norway, Canada, and the United States, and with the NATO Supreme Allied Commander. Each expressed concerns about sovereignty, access, and border protection; however, each suggested a strong preference for an Arctic strategy based on cooperation.

More recently, in April 2010, after 40 years of negotiations, Russia and Norway announced an agreement to end a long-standing undersea border dispute in the Barents Sea and Arctic Ocean. The agreement outlines the extent of each nation's Arctic territory.[21] While avoidance of military conflict cannot be assured, this committee's findings on potential conflict in the Arctic further support the 2005 national intelligence assessments that major military conflict in the Arctic region is not likely over the next 20 years.

As the Arctic region becomes more navigable, there is strong potential for a dramatic effect on global trade routes well beyond the Arctic. Although estimates on when the Arctic will become "ice-free" for purposes of safe commercial navigation range from 2013 to 2075,[22] two German commercial vessels did transit the Northern Sea route in the summer of 2009 with Russian icebreaker support.[23] China currently operates an icebreaking research vessel and is building a second, providing further evidence of increasing interests in the Arctic.[24] Very recently, Russia announced that it intends to send an oil tanker accompanied by an icebreaker from the White Sea to Japan via the Arctic route in the summer of 2010. The effort is believed by many to be an attempt by the Russian state-owned shipping company to demonstrate mastery of Arctic navigation.[25]

U.S. maritime forces must be prepared to play a part in this continuum of relationships in the Arctic—competition, cooperation, and conflict—by helping build maritime partnerships in the region and developing the requisite operational capabilities, as noted in previous chapters. In particular, combined operations, training, and planning between U.S. and Canadian maritime forces are going to be critical to protecting and promoting U.S. regional interests. In this and in other partnerships, the Navy and Coast Guard will be able to draw on established bilateral relationships and multilateral partnerships, such as the NATO alliance and the Arctic Council, but there will also be a need for new arrangements and agreements for Arctic maritime operations. It may be difficult, if not impos-

[21] See "Russia and Norway Reach Accord on Barents Sea," *New York Times*, April 27, 2010.

[22] As discussed in Chapter 2, throughout this report the term "ice-free" is used to mean that multiyear ice has nearly (or completely) disappeared; however, to date, in what is termed "ice-free" conditions, sufficient ice is present to remain a hazard to ordinary ships and routine marine operations. This committee suggests that 2030 is the approximate timing for ice-free summer months in the Arctic Ocean.

[23] See Andrew E. Kramer and Andrew C. Revkin, "Arctic Shortcut Beckons Shippers as Ice Thaws," *New York Times*, September 10, 2009.

[24] Linda Jacobson. 2010. "China Prepares for an Ice Free Arctic," *SIPRI Insights on Peace and Security*, March 1.

[25] See "Oil Tanker Titan Plans to Break the Ice on Arctic Route," *Financial Times*, April 13, 2010; available at http://www.ft.com/cms/s/0/7bcf96dc-4697-11df-9713-00144feab49a.html. Accessed June 4, 2010.

sible, for U.S. forces to develop these new arrangements and agreements if the United States fails to ratify the United Nations Convention on the Law of the Sea (UNCLOS). As discussed elsewhere, U.S. naval leadership should support ratification of UNCLOS.

FINDING 4.2: Although the likelihood of conflict in the Arctic is low, it cannot be ruled out, and competition in the region is a given. However, cooperation in the region should not be considered a given, even with close allies. Although there are mechanisms for bilateral and multilateral cooperation in the area, including the Arctic Council, these relationships and mechanisms are largely untested for emerging conditions. Additionally, with the ratification of UNCLOS, U.S. naval forces will be better positioned to conduct future naval operations and protect national security interests, especially in the Arctic.

RECOMMENDATION 4.2: The Chief of Naval Operations, working with the combatant commanders, the Commandant of the Coast Guard, and the Commandant of the Marine Corps, should build maritime partnerships in the Arctic region and encourage the United States to continue to identify and adopt policies and relationships in the Arctic that will build cooperation for new circumstances and minimize the risks of confrontation. (For example, naval leaders should pursue bilateral and multilateral training and exercising of U.S. naval personnel with partner nation personnel in maritime security, search and rescue, and HA/DR, and continue strong support of the U.S. efforts in the Arctic Council.) There should be no assumption that the geostrategic situation will take care of itself or that U.S. interests in the region are currently protected and promoted.

5

Climate-Change-Related Technical Issues Impacting U.S. Naval Operations

The technological infrastructure that supports naval operations is sophisticated, widely available, and reliable throughout the temperate and tropical oceans. It is, therefore, often taken for granted. However, the effects of climate change mandate that naval forces operate in areas that present challenges for the existing support systems and technologies. In particular, there is a high likelihood that a warming climate will increase the operational tempo in polar regions; consequently, the demands on navigation systems, communication systems, and nautical charts in polar regions will intensify. The initial increase in tempo will be driven by scientific and exploratory missions, especially so in the Arctic. As the degree of precision required by military combat operations can be more extreme than that required by peacetime operations, if tensions in the Arctic increase, the technical challenges will be multiplied. This chapter begins with an overview of naval navigation systems infrastructure and the resulting related climate change technical issues. The chapter then discusses communication systems performance in polar regions, followed by an examination of current nautical products and systems; also discussed is the critical role of ice characterization in operational safety in Arctic navigation. The chapter concludes by discussing climate-change-related antisubmarine warfare (ASW) impacts. The special challenges of submarine operations and ASW are of particular interest in the Arctic setting and are discussed separately.

CURRENT STATE OF NAVIGATION SYSTEMS INFRASTRUCTURE WITH RESPECT TO ARCTIC NAVIGATION

Navigation in the polar regions is challenging not only due to sea-ice and adverse weather conditions but also due to limitations of current navigation systems and infrastructure at high latitudes, which are degraded relative to performance in other regions of the world. This performance degradation affects surface, subsurface, and aircraft operations to varying degrees. Its significance to mission execution depends upon each mission's requirements for safe navigation in restricted water/airspace, precision localization and mapping, and the underlying accuracy of reference navigation charts.

Specifically, Global Positioning System (GPS) performance is degraded due to poor satellite geometry, larger ionospheric effects, and multipath interference. Similarly, the radio-navigation infrastructure that provides GPS corrections and/or position reference does not routinely extend to the polar regions. Magnetic heading becomes unstable and inertial navigation systems (INSs) suffer poor alignment above 70° north latitude due to the reduced effect of Earth's rotation. To prepare for expanded operations in the Arctic, the Navy should assess current military navigation system performance in polar regions and how it might inhibit operations. In addition, the Navy should seek to enhance the navigation infrastructure as necessary to prevent such limitations. Precision navigation is particularly crucial for combat military operations (precise tracking and targeting) and certain search and rescue operations.

GPS Performance Issues

Global Positioning System satellite orbit inclinations are at 55° to optimize performance in temperate and tropical regions of high activity. This results in low satellite elevation angles in polar areas, with approximately 45° being the highest satellite elevation angle possible at the poles. Data for satellites at low-elevation angles are more susceptible to ionospheric refraction and provide especially poor geometry for determination of a vertical position. The overall effect is minor for *surface* platform navigation, but it may be problematic for precision surveying and certain aircraft operations.

GPS coverage for surface navigation is only slightly degraded in the high latitudes (50 ft. horizontal precision has been demonstrated at the North Pole), but this accuracy is adequate for the navigational purposes of surface ships, aircraft, and submarines rising to the surface to obtain a navigation fix for undersea navigation systems. On the other hand, the degradation in vertical dilution of precision (VDOP) is much more significant and can result in altitude errors of up to 150 to 250 feet, which in turn could affect some Navy operations or system performance. Due to low-elevation angles, some attention to ensure clear lines of sight for the GPS antenna orientation is warranted for optimum performance.

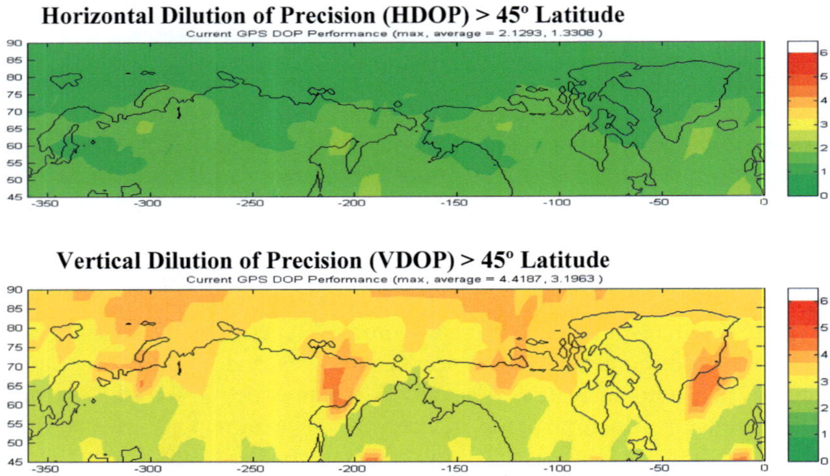

FIGURE 5.1 Horizontal and vertical dilution of precision above 45° latitude. Dilution-of-precision values of 1 to 2 are ideal. Values of 2 to 5 represent the minimum required for reliable navigation. Values above 5 represent low fix quality and may not be acceptable for certain military operations. SOURCE: Courtesy of The Boeing Company, Seal Beach, Calif.

Similarly, depending upon the placement of the antenna on a projectile versus the position of the GPS satellites on the horizon, guided munitions performance could also be adversely affected. Increased VDOP can also affect targeting when inaccurate height of target can transpose into horizontal error, depending upon the trajectory of the weapon. Figure 5.1 shows both horizontal dilution of precision (HDOP) and VDOP above 45° north latitude.[1]

Ionospheric Effects

Errors introduced by ionospheric delays are more pronounced in higher latitudes because of the reliance on low-elevation satellites. The ionosphere can affect GPS receivers by degrading the signal strength, in some cases causing code delay, phase advance, and loss of carrier lock. Additionally, irregularities in electron density, known as scintillation effects, can lead to significant phase and amplitude fluctuations in GPS signals as they pass through the ionosphere.

[1]For a more detailed discussion of GPS performance issues, see Dennis Milbert, 2009, "Improving Dilution of Precision," *GPS World,* November 1. Available at http://www.gpsworld.com/gnss-system/algorithms-methods/innovation-improving-dilution-precision-9100?page_id=3. Accessed August 2, 2010.

Military GPS receivers are dual frequency and can compensate for ionospheric delays; therefore, the real issue is the potential for GPS signal track loss. Because ionospheric compensation models are tuned for temperate regions, even dual-frequency receivers may experience more frequent GPS signal track loss.[2]

Multipath

GPS signals in the Arctic are subject to multipath effects where the GPS signal is reflected off the ocean and ice surfaces. This is due to the geometry caused by low-elevation satellites. These reflected signals can significantly affect the performance of GPS receivers, causing the systems to miscalculate position and speed. Depending upon the nature of the multipath effect, position error can persist for some time—as either a stable offset from truth, or an intermittent condition causing the GPS position and speed to fluctuate as the multipath signal comes and goes. In scenarios where many satellites are visible with good satellite geometry, the GPS receiver can discard "bad" values and multipath effects can be minimized. At higher latitudes, satellite geometry and visibility are already degraded; therefore, multipath effects are difficult to overcome.[3]

Inertial Navigation Systems

Many early generation navigation systems incorporate magnetic measurements for heading determinations. However, heading error can grow to unacceptable levels at very high latitudes because Earth's magnetic field lines are nearly vertical as one approaches the poles. More advanced integrated inertial navigation systems with GPS augmentation are advisable. It should be noted, however, that high-latitude operation poses a number of problems for INSs. Some of these problems are fundamental, such as the greatly reduced ability to determine azimuth by gyro-compassing at high latitudes, thus affecting self-calibration and alignment.[4]

Possible Solutions to Address Arctic Navigation Challenges

As explained above, GPS is an essential worldwide navigation aid that (due to ionospheric conditions and satellite geometry) provides slightly degraded service in the Arctic region, particularly increasing vertical navigation error with latitude

---

[2]For a more detailed discussion of ionospheric effects, see John A. Koubuchar, 1991, "Ionospheric Effects on GPS," *GPS World,* April. Available at http://gauss.gge.unb.ca/gpsworld/EarlyInnovationColumns/Innov.1991.04.pdf. Accessed June 4, 2010.

[3]For additional discussion of multipath effects, see "Sources of Error in GPS," April 19, 2009. Available at http://www.kowoma.de/en/gps/errors.htm. Accessed June 4, 2010.

[4]For additional discussion of inertial navigation systems, see "Inertial Navigation: Forty Years of Evolution." Available at http://www.imar-navigation.de/download/inertial_navigation_introduction.pdf. Accessed June 4, 2010.

increase. Meanwhile, other navigation sensors become more severely limited (inertial navigation systems do not align properly) or inoperable (magnetic heading error of 75 degrees is possible) at high latitudes. A combination of the effects of navigation sensor degradation could impact maritime operations, both from a charting and a naval forces' mission perspective.

Table 5.1 illustrates several alternatives to improve satellite-based navigation system performance at high latitudes. The table also considers the potential change to Navy user equipment, how well the solution would address the current

TABLE 5.1 Improvement Options for High-Latitude Satellite-Based Navigation

	Change in U.S. Navy User Equipment	Improve VDOP	Transmit Corrections	Comments
Add high-inclination satellites to GPS constellation	Low to medium impact	High impact	No	Additions beyond 32 satellites would require changes to receivers.
Use other GNSS constellations	High impact	Medium impact	No	Galileo-(56) Europe, GLONASS-(64.8)—Russian system satellite geometry better due to higher inclination.
MEO satellites using WAAS signals[a]	Medium impact	High impact	Yes	
Integrate GEO + LEO satellites for WAAS corrections[b]	Low impact (sidecar unit, backward compatible)	High impact	Yes	High-integrity GPS augmentation system (NRL R&D program)
Long-persistence UAVs	Medium impact	High impact	Yes	
Land-based beacons	Medium impact	Low impact	Yes	

NOTE: Acronyms are defined in Appendix B.

[a]Wide area augmentation system (WAAS) is a system of satellites and ground stations that provide GPS signal corrections, giving improved position accuracy.

[b]Low Earth orbit (LEO) satellites operate in orbits of around 100 km to 1,000 km above Earth's surface—much lower than traditional communications satellites—which bring them into frequent radio contact with ground stations. Because of their low orbits, a fleet of LEO satellites is required to maintain communications over a single point. In contrast, geostationary (GEO) satellites orbit at 35,786 km (22,236 miles) above Earth's equatorial plane, enabling the satellite to maintain the same position above Earth's surface at all times.

GPS VDOP challenges, and whether the improvement would accommodate the transmission of GPS error corrections to enable more precise navigation.

Options to improve satellite-based navigation fall into two categories: (1) using new satellites at higher orbit inclinations to cover the polar regions or (2) augmenting the GPS signal by transmitting corrections from either land-based beacons or a high-latitude overhead presence (such as other satellite systems or long-persistence unmanned aerial vehicles [UAVs]). Adding more satellites to the GPS constellation or using the satellite network put in place by other countries would require changes in user equipment and present logistics challenges that are likely unacceptable. The better solution involves transmitting GPS error corrections.

One such solution has been prototyped by the U.S. Navy's Naval Research Laboratory (NRL) High Integrity GPS Augmentation Demonstration Program—known more commonly as iGPS (see Figure 5.2). The program is developing techniques that enable faster acquisition time and augment GPS for military applications by exploiting the Iridium low Earth orbit (LEO) communications satellite system. Field tests have shown vast improvements in VDOP through use

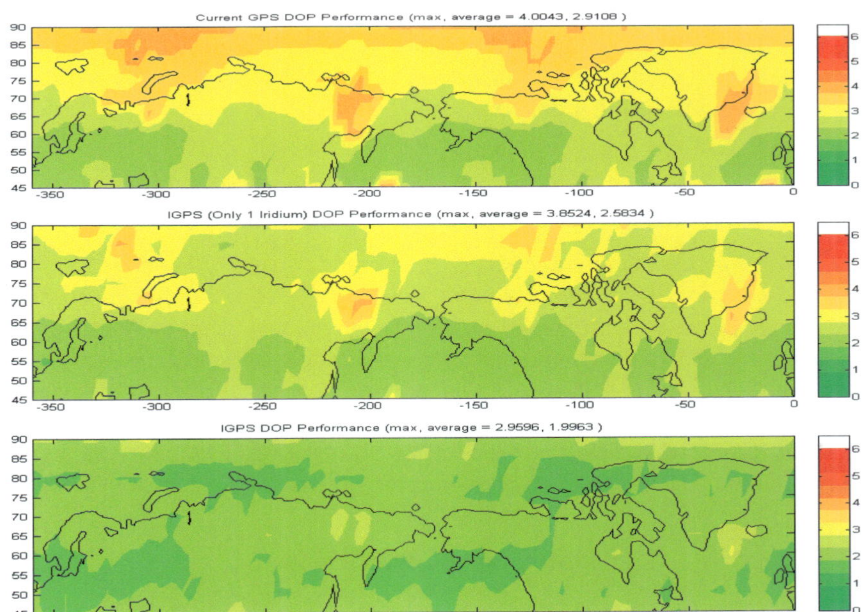

FIGURE 5.2 High-integrity Global Positioning System (GPS) augmentation system (Iridium-based) GPS vertical-dilution-of-precision improvement. SOURCE: Courtesy of The Boeing Company, Seal Beach, Calif.

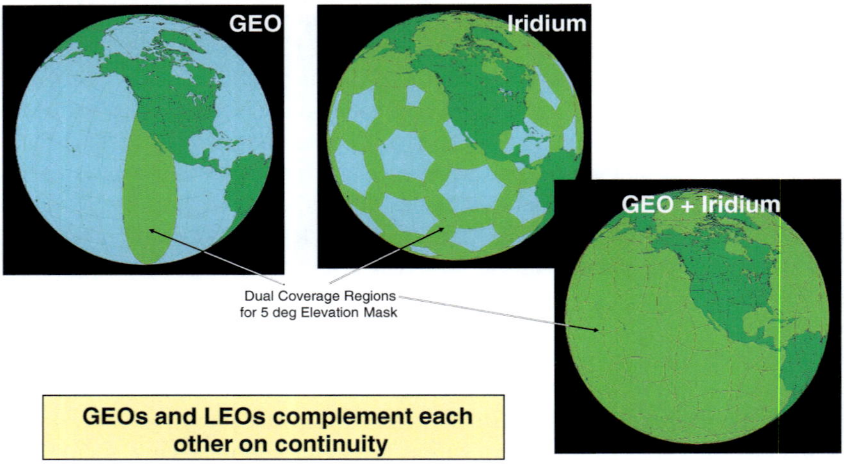

FIGURE 5.3 Wide area augmentation system data link coverage. SOURCE: Courtesy of the Federal Aviation Administration.

of the integrated geostationary/low Earth orbit (GEO/LEO) satellite network to provide expanded wide area augmentation system (WAAS) data link coverage for the polar regions (see Figure 5.3).

COMMUNICATION SYSTEMS INFRASTRUCTURE AND PERFORMANCE IN POLAR REGIONS

Although the impacts of climate change are not expected to directly impact radio frequency (RF) communication systems, there is a high likelihood that a warming climate will ultimately increase the operational tempo in Arctic regions and thus the demands on communication systems to operate in a familiar fashion and with performance standards similar to those that the naval forces have trained with and become accustomed to. Today's U.S. naval network-centric mobile communications architecture is designed around a mix of satellites for non-line-of-sight communications and line-of-sight (point-to-point) communication systems.

Line-of-Sight Communications

The line-of-sight communication systems employed by U.S. naval platforms provide horizon-limited local communications between Marine ground forces and among naval ships over useful ranges of 30 miles and between airborne assets up to hundreds of miles. These military communication and networking systems consist of multiple legacy systems and are characterized by the older Link

11 (high frequency [HF], very high frequency [VHF], and ultrahigh frequency [UHF] bands) and Link 16 (L-band frequency).[5] Throughout the world's oceans, but exacerbated in the Arctic, the HF and VHF bands are frequently degraded. Depending on asset location, HF and VHF bands can sometimes be of marginal naval platform use. This is due to both the scintillation of the ionosphere caused by solar wind electrons interacting with Earth's magnetosphere and the noise emanating from the galactic plane of the Milky Way.[6]

HF is known to be very sporadic and unreliable in the high-latitude environments due to the active ionosphere. Little can be done to mitigate these effects, and current operations typically suffer many hours of frequency outage. The HF systems are also expert operator-manpower-intensive and represent a skill set that is increasingly difficult to maintain. It is not uncommon for Coast Guard operations at higher latitudes to depend on low-elevation communications to GEO satellites—even if they require special positioning of the ship to gain favorable geometries—as opposed to struggling with HF systems.

Ionospheric disturbances of VHF voice and data communications are less intense than for HF bands, but they are still very problematic in the high Arctic (northward of 80° north latitude). The UHF and L-band frequencies are only slightly degraded while operating in the polar regions and are actively used by the Coast Guard in its deployments into the polar regions above Alaska.[7]

Over-the-Horizon Satellite Communications

The inherent limitations of line-of-sight communications systems have driven the military to adopt communication satellites to a bent-pipe relay system for over-the-horizon communication. The non-line-of-sight satellite communication systems are intended to provide communication for operational forces using a relay mode where two users are connected via an RF link relayed through a GEO satellite.[8] These GEO communications satellites are capable of both a one-to-one communication mode and a one-to-many, or broadcast, mode. They typically operate in higher microwave frequency regions.[9] These higher operating

[5]Link 11 operates at HF (10 to 30 MHz), VHF (120 to 225 MHz), and UHF (225 to 400 MHz); and Link 16 at L band (960 to 1215 MHz).

[6]See Norman Cohen and Kenneth Davies, 1994, *Radio Wave Propagation,* U.S. Space Environmental Laboratory, NOAA; available at http://www.swpc.noaa.gov/info/Radio.pdf. Accessed June 4, 2010.

[7]See United States Coast Guard Strategic Spectrum Plan, December 2007; available at http://www.ntia.doc.gov/osmhome/spectrumreform/Spectrum_Plans_2007/Coast%20Guard_Stategic_Spectrum_Plan_Nov2007.pdf. Accessed June 4, 2010. See also David N. Anderson, 2003, "Forecasting the Occurrence of Ionospheric Scintillation Activity in the Equatorial Ionosphere on a Day-to-Day Basis," *GPS Solutions,* Vol. 7, No. 3.

[8]See Executive Summary of the Commercial Satellite Communications (SATCOM) Report; available at http://www.fas.org/spp/military/docops/navy/commrept/index.html. Accessed June 4, 2010.

[9]See Geostationary Satellite History; available at http://www.geo-orbit.org/sizepgs/geodef.html. Accessed June 4, 2010.

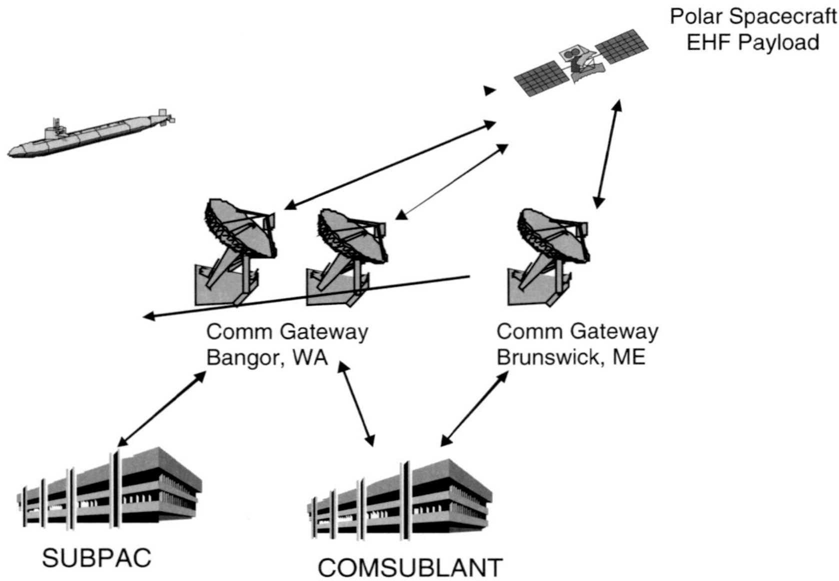

FIGURE 5.4 Current connectivity to submarine force worldwide with the addition of the Polar Interim Adjunct system. NOTE: Acronyms are defined in Appendix B.

frequencies are minimally impacted by Arctic environmental phenomena, but the geometry imposed by the high-latitude antenna coverage is the key limitation.

The two primary causes of over-the-horizon satellite communication degradation are the increased atmospheric RF losses due to increased path length at low antenna elevation angles and increased system noise due to antenna beam interception of the warm Earth as opposed to the cold background of space. Today's satellite systems are typically designed to function below 72°–65° latitude, depending on the time of day, due to the slight inclination (worst case 6°), which allows some visibility to extreme polar regions during portions of the day when spacecraft are at peak northern inclination.

The Submarine Satellite Information Exchange System (UHF) is also known to have limited reliability due to the orbital inclination residue of the GEO orbits.[10] (See Figure 5.4.) To support submarines operating at high latitudes above 65° north latitude, and as part of the submarine ice exercise (SUBICEX) 103 in

[10]GEO orbits are not stationary in inclination, and the satellites actually precess in a figure-eight pattern normal to the GEO plane. Satellite operators typically control this to keep within 6 degrees of the equatorial plane. See Submarine Satellite Master Plan; available at http://www.fas.org/man/dod-101/navy/docs/scmp/index.html. Accessed June 4, 2010.

January 2003, the Navy demonstrated limited communication coverage where visibility to GEO satellites is poor or impossible. Although limited, this communication capability is critical to submarine forces and allows consistent and reliable worldwide communications.

Polar Region Nautical Charting Products and Systems

General maritime operations and specialized contingency operations such as search and rescue in the polar regions are challenged in two key supporting areas: (1) nautical charting products that include coastal bathymetry, shoreline mapping, and coastal topography and (2) charts or maps that provide information on sea-ice conditions.

Accurate nautical charts in the polar regions are limited. In particular, nautical charts of the Alaska region show vast areas that have never been surveyed or have not been surveyed using modern instrumentation.[11] For example, Figure 5.5 shows the vintage National Oceanic and Atmospheric Administration (NOAA) charts for northern Alaska as of June 2008.[12] Many of the charts in those coastal areas are based on soundings from the 1940s or 1950s, with single-beam soundings, visual navigation, and surveys at small scale with line spacing of greater than 200 meters. The gaps extend to tidal data and tidal-current-prediction coverage.[13] These limitations in bathymetric soundings, coupled with shoreline data based on manual methods and poor topographic maps of near-coastal regions, are insufficient to support more widespread maritime operations.

Ice Characterization and Arctic Navigation

Knowledge of current ice conditions is crucial to safe maritime operations in the polar regions. The tri-agency Navy/NOAA/Coast Guard National Ice Center (NIC) produces various ice-related navigational products such as ice extent, daily ice edge and marginal ice zone, ice charts, and links to northern and southern

[11] As a typical example, see NOAA nautical chart 16549, Cold Bay and Approaches (Alaskan Peninsula); available at http://www.charts.noaa.gov/OnLineViewer/16549.shtml. Accessed August 2, 2010.

[12] "Maritime-Relevant Arctic Science at NOAA," briefing by John A. Calder, NOAA Climate Program Office, as contained in "Impact of Climate Change on Naval Operations in the Arctic," CNA annotated briefing CAB D0020034.A3/1REV April 2009 by Michael D. Bowes.

[13] NOAA is responsible for providing nautical charts of the Alaska region. The fundamental geospatial infrastructure that NOAA provides for the rest of the nation is lacking for Alaska and the Arctic, in particular. Alaska is the only state without digital shoreline imagery and elevation maps that meet nationally accepted standards. Also, the state's reference system has neither the density of control points to support submeter-level accuracies for surveying and positioning activities, nor vertical data coverage for the western half of the state to support the accurate determination of elevation heights. See CAPT James J. Fisher, USCG, Chief, Office of Policy Integration, Headquarters, "Waterways Management in the Arctic," presentation to the committee, September 25, 2009.

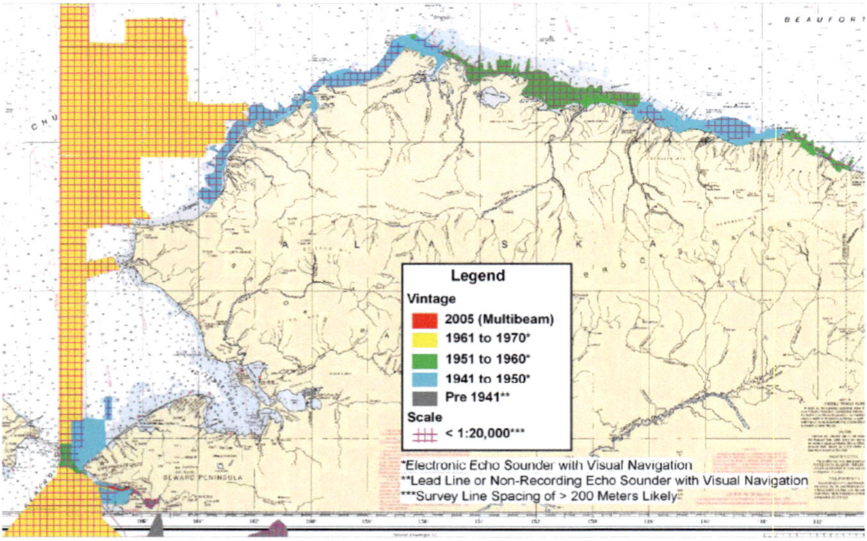

FIGURE 5.5 Vintage National Oceanic and Atmospheric Administration (NOAA) hydrography chart—North Slope (June 2008). SOURCE: Courtesy of the National Ice Center.

iceberg reports—all of which offer excellent information for operational planning. (The NIC also works with the Canadian Ice Service to jointly produce the North American Ice Service products.) The NIC ocean ice and iceberg products are based principally on satellite passive microwave and synthetic aperture radar (SAR) data. The ice charts include a rough characterization of ice thickness and types where available.

Ice conditions in the marginal ice zone can change very quickly, generally as a result of new ice formation and breakup, the latter coming principally from current year ice, known as seasonal ice. Since seasonal ice is more prone to breaking up and creating dangerously dynamic ice floe conditions than is multiyear ice, augmentation of sea-ice coverage charts with ice thickness estimates would be very useful for maritime operations safety—including surface, subsurface, and certain air operations. Similarly, near-real-time characterization of ice concentration and features in the marginal ice zones would greatly enhance operational safety. However, distinguishing between seasonal and multiyear ice is challenging via remote sensing, as new ice is generally less than 2 meters thick and old ice 3 meters or more. Therefore, ice thickness measurements must be accurate to within 0.5 to 1.0 meter for this purpose. Even then the classification between seasonal and multiyear ice is not straightforward.

As mentioned above, the thickness and temporal and spatial distribution of sea ice can dramatically affect navigation decisions. Unfortunately, it is difficult

TABLE 5.2 Currently Operating Spaceborne Synthetic Aperture Radars (SARs)

SAR	Launch Date	Frequency	Polarization	Resolution
ERS-2	1995	C-band	VV	25 m
Radarsat-1	1995	C-band	HH	25 to 50 m
Envisat	2002	C-band	HH.VV, VV/HH, HV/HH, VH/VV	30 to 1,000 m
ALOS PALSAR	2006	L-band	Full-polarization	7 to 88
TerraSAR-X	2007	X-band	Full-polarization	3
Radarsat-2	2007	C-band	Full-polarization	3 to 100
Cosmos SkyMed	2007	C-band	Full-polarization	3

to directly measure ice thickness from space. Lidars, like the ones on IceSat and all-weather radar altimeters from SeaSat (1978) to the recently launched CryoSat, can measure sea-ice freeboard.[14] Freeboard, the distance from the water line to the top of the ice, is correlated with total thickness. The limitation of altimeters is that they sample only single points at the nadir along the satellite ground tracks. Altimeters can create a very narrow grid of measurements over many months or revisit the same spots more frequently, albeit with a thinner grid. In addition, the presence of snow or ice crystals on the surface of the ice biases the inference of thickness as there is a 10:1 amplification of the bias.

The last few years have proven to be a golden age in spaceborne synthetic aperture radars that create radar cross-section images. Table 5.2 lists the currently operating spaceborne SARs. In 2012, the European Space Agency (ESA) plans to launch Sentennial-1, which will provide C-band SAR imagery on an operational basis. The United States hopes to launch the L-band DESDYNI (deformation, ecosystem structure, and dynamics of ice) SAR[15] within a decade. Both of these SARs can operate in a variety of modes from narrow to wide swath and at a variety of polarizations or even multiple polarizations.

SAR imagery can be used to find ice-free areas and to infer ice age from both absolute radar cross section and image texture. Ice age/type estimation is aided when the same areas are imaged at different frequencies and polarizations. Ice type is correlated, though imperfectly, to ice thickness. SARs can also operate in cross track interferometry mode, which can make vertical height measurements of 1 meter resolution, depending on a number of system factors. DESDYNI will be an interferometric SAR. In addition, though limited by clouds, optical and infrared imagery from NOAA weather satellites and NASA's Moderate Resolution Imaging Spectrometer (MODIS) instrument help in estimating sea-ice extent.

[14]A lidar (light detection and ranging) is a remote sensing system used to collect topographic data.

[15]The DESDYNI satellites, sponsored by NASA, will be a dedicated U.S. interferometric SAR and lidar mission optimized for studying hazards and global environmental change. More information is available at http://desdyni.jpl.nasa.gov/. Accessed June 4, 2010.

No single instrument or instrument type yields high-resolution, timely measurements of ice thickness. Lidars and altimeters make the most direct measurement of ice thickness, but they are thinly spread. SARs and other imaging instruments can measure over relatively large areas at high resolution, but they provide indirect inference of ice thickness except when they are in an interferometry mode.

It is clear that the best approach is to combine the measurements, perhaps linked by a sea-ice model, using the imaging sensors to create ice thickness images that are tied to fiducial measurements by lidars and radars. The optimum combination of instruments, operating frequencies, polarizations, operating mode, and orbit patterns to provide these measurements is an area where additional research should be applied and a proof-of-principle demonstration should be performed. It may turn out that a modest modification of operating modes or the launch of key sensors at key times would have significant impact on the timeliness and accuracy of operational ice thickness estimates.

Additionally, analysts play a key role in translation of sea-ice data from multiple sources—text or verbal ice reports, in addition to remote sensing data sources—into useful products to support Arctic navigation. The work of the NIC analyst operations will continue to grow in importance as the operational tempo in the Arctic increases. In the committee's opinion, real-time ice characterization and maps in emergent Arctic routes are needed now to avoid emergencies that would require Navy or Coast Guard involvement, and to support such involvement if and when it happens.

FINDING 5.1: U.S. military navigation and communications systems have been optimized to support operations in non-polar regions. Likewise, data on terrain elevation and bathymetry to support military operations and nautical charting are of low resolution and sparse in the Arctic. Moreover, while accurate ice coverage charts are available to guide surface navigation, reliable real-time ice characterization and maps in emergent Arctic transit routes are not. The combined effect of degraded navigation, communications, and charting systems could impact safe operations and reduce the performance of military systems in the polar regions.

RECOMMENDATION 5.1: The Assistant Secretary of the Navy for Research, Development, and Acquisition should increase research and development efforts at the Office of Naval Research and the Naval Research Laboratory to address the operational shortfalls of existing and planned navigation, communications, and charting systems, leveraging both local and global augmentation technologies. In conjunction with the National Oceanic and Atmospheric Administration, the Department of the Navy should increase priority for extending modern navigation, communications, and charting coverage to include the Arctic region.

ANTISUBMARINE WARFARE

Global Antisubmarine Warfare Operations

There are no significant first-order effects from climate change on U.S. antisubmarine warfare capabilities. A robust infrastructure that collects, analyzes, and distributes oceanographic data essential to ASW effectiveness is in place and covers active submarine operating areas adequately. Climate change will, however, mandate that submarine and ASW operations become more robust in the Arctic Ocean, where essential data are sparse or nonexistent in both special and temporal senses. Moreover, as potential adversarial submarines have become acoustically more quiet, ASW operations have evolved away from a pure submarine-on-submarine mission to a cooperative, coordinated mission involving fixed and mobile sensors and surface, subsurface, and air platforms.

This extensive and deployable ASW infrastructure that supports the principal nuclear-powered attack submarine (SSN) hunter platforms is generally deployed in the temperate oceans, but it would be challenged to operate in the Arctic. As well, the supporting tactical oceanographic data collection, analysis, and distribution system does not extend to the Arctic. Additional support infrastructure must be established or restored to enable more effective ASW operations in that region, which will become an inevitable national imperative.

Ocean acoustics are fundamental to submarine operations and antisubmarine warfare. The speed of sound in seawater is a function of pressure (depth), temperature, and salinity. Acoustic waves reflect off the sea surface and seafloor boundaries, and seawater absorbs acoustical energy at a rate proportional to frequency squared. There are two net effects of these properties upon ocean acoustics, which can be summarized as follows: (1) the refractive properties can lead to a sound fixing and ranging (SOFAR) duct that traps energy and leads to very-long-range propagation of signals at low frequencies, and (2) the combination of boundary losses and absorption losses leads to an optimal frequency for efficient sound propagation.[16] These effects, plus the ambient noise environment and capabilities of the sonar system, determine the performance (for example, detection range) of an ASW system. For concerns of this report, ocean climatology impacts both of these major effects as well as the ambient noise.

Refraction (and the resulting SOFAR duct) is a phenomenon that can lead to long-range detection of submarines. In temperate oceans, the SOFAR duct is typically at a depth of 1 km, but as one goes to higher latitudes and colder water, it gradually migrates to the surface, which is the case in the Arctic Ocean. The dominant trade-off for the depth of the SOFAR duct is between hydrostatic pres-

---

[16]The sound speed in the oceans is a function of depth. In deep water the opposing effects of warm water at the surface and higher density at depth lead to a minimum in the sound speed. Since sound will always bend, or refract, toward a minimum, this leads to a duct trapping the acoustic power and very low loss propagation.

sure and temperature, with salinity playing a lesser role. Salinity is more important for determining absorption rates and the optimal frequency.

There is nothing novel about these processes. The U.S. Navy and other world navies have invested large sums to acquire field measurements of temperature and salinity, as well as bathymetry, to produce climatological "atlases" available as a function of time of year for strategically and tactically important regions throughout the world. These atlases are maintained by the Naval Oceanographic Office (NAVOCEANO) by collecting the data produced daily through expendable bathythermographs (XBTs) and expendable conductivity temperature and depth (XCTD) from ships at sea. The atlases are critical in ASW detecting, localizing, and tracking potential adversarial submarines.

In addition, NAVOCEANO maintains a set of sophisticated prediction codes that forecast oceanographic conditions for use by the fleet.[17] These "nowcasting" and forecasting models also merge archival data and in situ data in an optimum statistical manner accounting for currents, winds, historic sound speed profiles, and the accuracy of the in situ data. These are available to U.S. Navy ships on a 24-hours-a-day, 7-days-a-week basis.

The major issue here is that ocean temperature and salinity are highly spatially and temporally variable, so an ongoing and expensive measurement campaign is needed to keep these atlases up to date.[18] It would be comforting to assume that climate-induced ocean changes will be slow, and that the impact on current data atlases will be minimal; however, not enough is known about climate change to be assured of these assumptions.

One can also make similar claims about the ambient noise environment. Ocean noise is primarily a function of shipping density, ice noise, and animal/sea life vocalizations. It is known that shipping noise is increasing and that wind stress will change as the climate changes. Simply not enough is known about potential climate-change-related impacts on marine animals to make any predictions related to the noise environment, except that temperature and salinity changes will almost certainly lead to changes of habitat.

Arctic Antisubmarine Warfare Operations

The reduction in Arctic sea ice and the increased exploration accessibility to potential natural resources has already led to the Arctic nations posting overlapping and disputed claims of territory, as discussed in earlier chapters of this

[17]The prediction codes model just the water column and not the atmosphere. They have a 7-day duration for providing useful predictive information. Codes that couple the air-sea boundary would extend prediction durations.

[18]Atlases are typically maintained on a monthly basis for temporal variability. Spatial variability is a function of the survey density.

report.[19] The claim by Russia of virtually the entire basin from its Siberian coast to the North Pole is the most audacious (dramatized by placing a titanium Russian flag beneath the North Pole on the Lomonosov Ridge). The basis of Russia's claim is that the Lomonosov Ridge is a continental fragment split from Siberia due to seafloor spreading and hence part of Russia's continental margin even though it is completely submerged. This is the major, but one of four, disputes among the Arctic nations.[20] Others involve the disputes over status of the Northwest Passage and Northern Sea route and involve questions such as whether the Northwest Passage is similar to an extended strait between two seas through which, therefore, innocent passage is assured. Also, is the Northern Sea route, while within Russia's exclusive economic zone (EEZ), subject to rules governing innocent passage?

Where does ASW enter? Submarines are a primary U.S. Navy asset that can assert the national will in an international "hot war" dispute in the Arctic. It is hard to consider a scenario escalating to this level, but a credible ASW threat in the Arctic could be needed as part of negotiations. Even during peacetime, many countries attempt to know the location of submarines of potential adversaries. Sometimes this includes submarine-on-submarine events in which both could be doing ASW on the other. One can easily envision a peacetime situation in which countries deploy submarines as a statement of territorial interest and capability, just as Russian bombers and icebreakers have done recently within their newly claimed territory. This might lead to ASW-like operations in which everything is done except armed engagement. One does not have to have a "hot war" for ASW missions to take place.

Arctic Antisubmarine Warfare

Arctic Ocean ASW is especially sensitive to the issues outlined above. Currently, virtually all knowledge of Arctic climatology is from submarine transits. While there have been many transits, they do not come close to the number needed for a high-resolution atlas. The most extensive and up-to-date data are for bathymetry, since the seafloor changes slowly; however, temperature and salinity data are largely dependent upon historical data gathered at ice camps primarily by the former Soviet Union (FSU) and the United States. While more data are becoming available as icebreakers are sent for scientific use into the Arctic, U.S. naval forces still are far short of the fidelity of temperate ocean atlases. Consequently, the aspects of Arctic Ocean ASW dependent upon the environment are already

[19] Jon D. Carlson, Christopher Hubach, Joseph Long, Kellen Minteer, and Shane Young. 2009. "The Scramble for the Arctic: The United Nations Convention on the Law of the Sea and Extending National Seabed Claims," paper presented at the annual meeting of the Midwest Political Science Association 67th Annual National Conference, April 2, The Palmer House Hilton, Chicago, Ill.; available at http://www.allacademic.com/meta/p363540_index.html. Accessed June 4, 2010.

[20] See Ronald O'Rourke, 2010, *Changes in the Arctic: Background and Issues for Congress,* March 30, Congressional Research Service Report for Congress, Washington, D.C., pp. 7-12.

data poor, and this deficit will certainly increase with climate change. Most of the models for predicting climate change indicate the high latitudes will respond the earliest. This is being observed with the retreat of the seasonal Arctic sea ice.[21] In summary, there is a sparse data set, and the committee notes that it will need to be updated more frequently because of the more rapid changes at the polar latitudes.

The United States had a robust Arctic research program corresponding to the era when the FSU conducted extensive operations there. This continued until the end of the Cold War, when the Office of Naval Research (ONR) and most other Navy program offices closed their Arctic operations. The Science Ice Exercise Program (SCICEX) cruises in which the U.S. Navy sent an SSN on transits across the Arctic for the scientific community continued until 2000, when the last of the SSN 637 class strengthened for Arctic surfacing was retired. Since then there have been some SSN Arctic transits between the Atlantic and the Pacific, but not for scientific purposes. The U.S. Navy research program in the Arctic has atrophied to the point that there is no infrastructure to support it. The National Science Foundation (NSF) is the current primary U.S. federal source of support for Arctic science and technology.

The SSN transits, nevertheless, have been extremely valuable for surveying the Arctic Ocean bathymetry. The data from these transits have been compiled by NAVOCEANO with those from other sources such as icebreakers and ice camps. Nevertheless, these data are still sparse and certainly not suitable for routine navigation, especially near the shelf break. The sparse bathymetric charts lead to the challenge of SSNs avoiding undiscovered seamounts, which are still being identified.[22]

Detection, Classification, Localization, and Tracking in the Arctic

ASW is often divided according to the tasks of detection, classification, localization, and tracking. The committee examined how the Arctic and impact of climate change can affect these tasks.

Detection is fundamentally a signal-to-noise issue, so the transmission loss from a target, or source of acoustic power, is strongly influenced by the refractions and reflection taken by the path before being received. Most of the time,

[21] In addition to the retreat of the ice cap, a warming of 0.4°C for the Atlantic intermediate water mass north of Greenland has been measured by acoustic tomography (P.N. Mikhalevsky, A.B. Baggeroer, A. Gavrilov, and M. Slavinsky, 1995, "Experiment Tests Use of Acoustics to Monitor Temperature and Ice in Arctic Ocean," *EOS, Transactions American Geophysical Union,* Vol. 76, No. 27, p. 265) and directly by SCICEX transits.

[22] For example, the Coast Guard icebreaker *Healy* discovered a seamount during a cruise in 1989. While it shoaled to 3,200 meters from 5,000 meters and was not a navigation hazard, it was in the middle of the Canada Abyssal Plain and completely unexpected geologically. See *Arctic Mapping and the Law of the Sea;* available at http://arctic-healy-baker-2008.blogspot.com/2009/09/new-seamount.html. Accessed August 2, 2010.

both the source and the receiver are within the upper part of the water column where the halocline and thermocline exist.[23] The propagation is described as refracted-surface reflected (RSR). Acoustical energy reflects off the ice canopy and is refracted by the sound speed gradient at depth. Climate warming will lead to more freshwater from sea ice, glacier runoff, and the northern rivers, which will affect salinity distribution. In addition, surface heat will warm the upper waters, increasing the speed of sound. This will weaken the surface duct and cause sound to refract away from the water or ice surface, leading to changes in transmission loss and detection levels depending upon where the source and receiver are in the water column.

For ASW purposes, classification is the task of determining the source of a detected sound (for example, a submarine, commercial shipping, marine life, or even ice movement or wind). Outside the Arctic, interference from commercial shipping leads to a lot of "clutter" on displays. This makes identifying a target difficult. There are currently few ships in the Arctic, but warming may lead to increased maritime trade, more ships, and more difficulty in classification, although this will probably not reach the level of difficulty seen in sea lines of communication in the temperate oceans.[24] Biologic noise, as well as that due to ice activity in the marginal ice zone, is very high and can aggravate classification efforts.

The impact of ambient noise for Arctic ASW leads to some interesting questions and speculations. As noted above, in temperate oceans the noise in the interesting ASW bands is dominated by shipping and rain/wind noise. In the Arctic, with an ice canopy, there is virtually no shipping and the rain/wind is isolated by the ice cap. The dominant noise is the ice grinding against itself and the seabed; this is especially loud in the marginal ice zone, the transition from open water to the ice pack. Arctic ambient noise can be very quiet or very loud. If shipping increases, the ambient noise will likely increase; however, it is uncertain whether it will be a factor in all but the quietest days for ASW. It is possible that the newly opened part of the Arctic Ocean will be similar to sections of the southern hemisphere, which is noted for low ambient noise.

Localization and tracking are the tasks of determining the range, bearing, speed, and course of a submarine. While these tasks certainly depend upon signal-to-noise ratios as well as interferences, the issues are not different in the polar regions from those of temperate waters. To the first order, these efforts should not depend upon Arctic Ocean warming.

Another important component of ASW is weapons performance, usually a torpedo. Torpedo operation would not be materially affected by changes in salin-

[23]The halocline is a narrow vertical gradient of salinity where meltwater from the ice decreases the salinity of the water near the surface. The thermocline is a narrow vertical gradient of temperature, usually within 300 meters of the surface. These combine to form a pyncoline, or density gradient, and a more focused surface duct beyond that formed by the overall upward refracting profile.

[24]Sea lines of communication are the primary shipping routes between ports.

ity and its impact on absorption; however, the additional complications of the stratification of the halocline and thermocline, plus scattering from the underside of the ice, will cause tracking problems for the homing system on a torpedo. By far the biggest problem for a torpedo is reverberation from the ice canopy, so more open water implies larger regions where the torpedo would not suffer performance degradation from the ice canopy. The United States already has a program for assessing the performance of torpedoes under the ice.

ASW is best done by submarine, but the submarine does rely upon an infrastructure to provide a set of cues to help vector it to a target. Maritime patrol aircraft (P3s and now P8s) drop sonobuoys to assist in a prosecution, but they will be disadvantaged because of the long ranges from an airfield and the existing ice canopy. These systems could provide surveillance at important choke points and yield valuable cues. In summary, ASW is in many ways a team effort needing the cooperation of many systems to cue an attacking submarine to a target. If these supporting systems or infrastructures are not available, ASW reverts to submarine-on-submarine engagement that disadvantages the pursuer.

As mentioned earlier, the Virginia-class submarines were not constructed to penetrate thick ice. Locations for surfacing need to be carefully checked to make sure the ice is thin enough for these submarines to penetrate without damage. This implies that there is a capability for finding regions of thin ice for surfacing opportunities, and this needs to be put in place based either on data from the upward looking sonar or by somehow transmitting satellite reconnaissance information.

Finally, many of the personnel who had the skills to operate in the Arctic have gradually retired or otherwise left the ASW community. There is no formal program for training to develop the Arctic skill sets needed.

FINDING 5.2a: Arctic ASW is difficult because of the complications of the environment—the submarine and a source are typically in the section of the sound fixing and ranging (SOFAR) channel that has the most variability in sound speed. While the bathymetry does not change, it is poorly sampled in terms of both coverage and accuracy, and the ice canopy prevents routine submarine surfacing for emergencies and satellite communication. In addition, an ice cover scatters sound, which limits detection and torpedo performance.

FINDING 5.2b: The United States had an Arctic research program during the Cold War that has essentially ceased. Moreover, there is no infrastructure to support antisubmarine warfare (ASW) in the Arctic. While there are no significant ASW activities now in the Arctic, U.S. naval forces need to be prepared to operate there safely. The United States' diminished Arctic research program and capabilities from what existed during the Cold War—plus the need for even better performance from its ASW systems—put U.S. naval forces' ability to operate as needed in the Arctic at risk if the United States does not keep pace with the capabilities of other Arctic nations, especially Russia with its extensive claims of Arctic sovereignty, as well as with non-Arctic nations, such as China.

RECOMMENDATION 5.2: Given that climate change may drive the U.S. naval forces to conduct antisubmarine warfare (ASW) operations in the Arctic, the Department of the Navy should increase its submarine Arctic presence for training purposes, extend its supporting ASW oceanographic data infrastructure to the Arctic Ocean, and begin to conduct multiplatform ASW training exercises in the Arctic. Specifically, this should include:

- Increased research for Arctic passive and active sonars;
- Long-range planning to install facilities that support Arctic ASW, such as refurbishing and expanding the fixed array systems;
- Planning for aircraft support from the new P8;
- Development of high-latitude communications systems for relaying tactical and environmental data;
- Identifying ports for emergencies; and
- Incorporation of a more robust under-ice capability on Virginia-class submarines.

6

Future Research and Development Needs

INTRODUCTION

The domain of U.S. maritime forces—namely, coastal waters, the high seas, and shore-based facilities—is projected in Intergovernmental Panel on Climate Change *Fourth Assessment Report* (IPCC AR4) scenarios to be affected as climate is increasingly altered by rising atmospheric concentrations of greenhouse gases. Temperature data provide evidence for warming of the atmosphere and ocean over the past several decades. Estimates show that about 90 percent of the heat now accumulating in Earth's climate system is being stored in the ocean.[1] This fact is often underappreciated, however. Within the past few decades, direct measurements show that ocean temperatures across vast regions of the high seas are now elevated at depths of a thousand or more meters relative to the first half of the 20th century. Some of the incremental heat being retained by the enhanced greenhouse effect is affecting the ocean in another way: the melting of sea ice is changing maritime access in the Arctic and the melting of land ice is contributing to sea-level rise, as discussed in prior chapters.

The mission of U.S. naval forces requires knowledge of ocean conditions. Historic databases, which constitute ocean climatology, are an essential component of forecasting and predicting systems used in several aspects of naval operations. It is clear that projected and reported climate change now under way is compromising the value of established climatologies that have been used for these purposes.

[1]S. Levitus, J.I. Antonov, T.P. Boyer, R.A. Locarnini, H.E. Garcia, and A.V. Mishonov. 2009. "Global Ocean Heat Content 1955–2008 in Light of Recently Revealed Instrumentation Problems," *Geophysical Research Letters,* Vol. 36, L07608.

This chapter provides an assessment of future climate change research and development (R&D) needs related to U.S. naval operations. Such needs are for supporting naval tactical operations and for providing improved data for future U.S. naval planning. The committee's examination of topics for future R&D emphasis focuses on those areas in which the naval forces have particular interests that might not likely be met in the near term by other groups pursuing climate-related research. It is specifically recommended that the Navy address R&D issues related to climate observations, climate modeling, and sea-level rise, as well as needs unique to the Arctic. Other important climate research questions have implications for U.S. naval forces, but it is expected that many of these will be pursued in the course of ongoing and/or planned U.S. and international scientific programs. For example, there is a set of important questions related to the consequences of a decrease in ocean pH resulting from the increasing ocean absorption of carbon dioxide from the atmosphere. This is an area of basic research that the general scientific community is vigorously pursuing. Some aspects of ocean acidification that might be of special importance to the Navy—such as the potential effects of a pH decrease on sound absorption—are still under debate.[2] The Navy should continue to monitor the research in ocean acidification closely, as the results may hold potential important implications for ocean acoustics critical to U.S. naval operations. The committee concluded that formulation of specific recommendations in this area would be premature.

Improved understanding of how climate is changing will surely point to new research areas of particular importance for U.S. naval forces. As a nation, we need to be prepared for surprises.

GLOBAL OBSERVATIONS, SCIENTIFIC ANALYSIS, AND MODELING IN SUPPORT OF NAVY R&D REQUIREMENTS

Naval operations demand environmental information in the form of observations, model-based analysis products, and model forecasts for navigation, communication, general fleet support, antisubmarine warfare (ASW), and search and rescue. The Navy has long had programs in place to collect ocean and marine meteorological data for these purposes. It also has well-established weather,

[2]Oceanographers Tatiana Ilyina and Richard Zeebe of the School of Ocean and Earth Science and Technology at the University of Hawaii at Manoa, together with Peter Brewer of the Monterey Bay Aquarium Research Institute, have hypothesized that seawater sound absorption will drop by up to 70 percent during this century. The scientists have examined the effects of human-made carbon dioxide under business-as-usual emissions and provide projections of the magnitude, timescale, and regional extent of changes in underwater acoustics resulting from ocean acidification. These changes are projected to be associated with the fact that low-frequency sound absorption depends on the concentration of dissolved chemicals such as boric acid, which in turn depends on seawater pH. These researchers also explained that further research is needed to address key questions in this area. See "Man Made Carbon Dioxide Affects Ocean Acoustics," *Science News,* December 22, 2009.

ocean, and sea-ice modeling and forecasting capabilities. The Navy's R&D efforts are intended to infuse new model, computing, and observational technologies into operational capabilities.

At present, almost all data collection and modeling efforts within the Navy have a marine focus for tactical purposes. It is anticipated that requirements for such tactical scale observations will continue and that they will be an integral part of naval operations in the future. Here the committee addresses the related but distinct requirements for global-scale observations and modeling as part of a naval R&D climate change risk management strategy over the next 30 years. This R&D effort is intended to provide the information necessary for enhancing the U.S. Navy's maritime domain awareness and for reducing uncertainties in seasonal to decadal timescale forecasts that guide long-range Navy planning.

Projected effects of climate change suggest it will alter the physical environment in which the Navy operates in the coming decades. Warming ocean and land temperatures, rising sea levels, disappearing Arctic sea ice, shrinking glaciers and ice sheets, shifts in rainfall patterns, and changes in storm frequency, intensity, and spatial distribution are among the projected manifestations of climate change. The implications of these changes are such that the climatological databases that the U.S. naval forces have used in the past may no longer be valid in the future.

Against this backdrop, U.S. naval forces will become even more dependent in the future on observations, analysis products, and forecasts of the global environment to carry out its mission. The U.S. naval forces' needs will be largely focused on the maritime environment as in the past; because of their humanitarian assistance/disaster relief (HA/DR) mission and shore-based facilities, however, naval forces will also require information on evolving environmental conditions in continental regions where vulnerabilities to climate change are greatest. There will also be regions, like the Arctic, that require special attention because of the unique mix of environmental, societal, and national security issues that they present.

Current Status of Global Ocean Observations

To carry out its mission, the U.S. Navy needs many ocean, atmosphere, cryosphere, and land measurements. Key parameters that need to be measured in the marine environment to support naval operations include temperature, salinity, ocean currents, surface waves, coastal sea level, sound speed, ambient noise, and, in polar regions, sea-ice extent and thickness. Marine meteorological measurements are also needed for winds, air temperature, pressure, relative humidity, precipitation, and other parameters. At present, these data are required to support fleet operations on tactical timescales and space scales. They are used primarily for describing current conditions and for forecasting evolving conditions in the oceans and the atmosphere on timescales of about a few days to a week. The discussions in this section extend beyond the tactical scale to address global sustained ocean observing systems for climate and how they can address future mission goals of the Navy.

Currently, global measurements in the marine environment come from a mix of Earth observing satellites and in situ sensors as part of the Global Ocean Observing System (GOOS) and the Global Climate Observing System (GCOS). GOOS and GCOS represent international coordination efforts sponsored by the International Oceanographic Commission (IOC), World Meteorological Organization (WMO), and International Council for Science (ICSU) dating back to the early 1990s. The United States is the single largest contributor to these programs, with most U.S.-sponsored measurements made by civilian agencies such as NASA-, National Oceanic and Atmospheric Administration (NOAA)-, and National Science Foundation (NSF)-sponsored researchers. There is a premium placed on real-time and near-real-time data availability as it enables timely and routine monitoring while providing data for weather and climate forecast model initialization.

Elements of this observing system include NOAA and NASA satellites that are a critical source of information on the global- and regional-scale ocean sea-surface-temperature warming trends. Sea-level rise over the past 15 years has been tracked by NASA and European Space Agency (ESA) altimeter missions. In situ components include moored and drifting buoy arrays, such as the Argo float program.[3] These and other satellite and in situ measurement efforts benefit Navy operational weather and ocean forecasting at the Navy's Fleet Numerical Meteorology and Oceanography Center (FNMOC), and the Commander, Naval Meteorology and Oceanography Command (CNMOC), by providing key data sets for model initialization and verification. A recent technical conference highlighted the history, status, and plans for further development of the ocean observing system, with nearly 100 community white papers published on the subject as part of its proceedings.[4]

The Navy and the Department of Defense (DOD) have measurement assets geared toward addressing the needs of their specific mission sets, but which contribute to the ocean observing system. These include, for example, the polar-orbiting satellites of the Defense Meteorological Satellite Program (DMSP), from which much of the data is declassified and publicly available. DMSP data, together with data from NOAA and NASA satellites, have provided the clearest evidence for diminished Arctic sea ice in summer.

---

[3]Argo is a global array of 3,000 free-drifting profiling floats that measures the temperature and salinity of the upper 2,000 m of the ocean, allowing continuous monitoring of the temperature, salinity, and velocity of the upper ocean. All Argo data are relayed and made publicly available within hours after collection.

[4]For a summary of OceanObs09, see D.E. Harrison and David M. Legler, 2009, "Saltier, Hotter, More Acidic, and Less Diverse? Observing the Future Ocean," *EOS, Transactions American Geophysical Union,* Vol. 91, No. 3, p. 23. See also a compilation of meeting papers at http://www.oceanobs09.net/. Accessed June 4, 2010.

There are classified Navy and DOD measurements and assets whose access is restricted for reasons of national security. These could make significant contributions to GOOS if more open access were provided. Classified historical data would be invaluable in developing observational baselines for gauging current and future climate change. The committee believes that additional such data can be released that would not be harmful to national security, nor would it compromise other sensitive information concerning the types of measurements, methods, equipment, positional requirements, and so on. As one example, release of images of Arctic sea ice from 1999 to the present as part of the Measurements of Earth Data for Environmental Analysis (MEDEA) Program is providing unique and fundamentally new information on the loss of Arctic sea ice, which is largely attributable to climate change.[5] In another example, the release of Arctic sea-ice draft data derived from submarine upward looking sonar from the Navy's Arctic Submarine Laboratory provides a critical long-term estimate of sea-ice thickness since 1975.[6]

Despite the widespread interest nationally and internationally in developing a global ocean observing system for climate, there are significant challenges that limit progress in addressing some of the most pressing problems. For example, it is not possible yet to routinely measure properties in the deep ocean below 2,000 meters from autonomous platforms. These measurements are needed to accurately document the ocean's storage of heat. Similarly, western boundary currents like the Gulf Stream are critical conduits for meridional transport of oceanic mass, heat, and salt to the poles. Current data collection technologies make it difficult and expensive to consistently gather long-term measurements in these high-velocities regions. The Arctic, discussed in more detail later, is a particularly challenging environment for sustained ocean observing systems because of its extreme cold, remoteness, and ice cover most of the year, even though half of the Arctic basin lies within the exclusive economic zones (EEZs) of rim nations.

From a climate perspective, data records have greatest value when they are multiyear to multi-decadal and longer in length, continuous in time, and with sufficient meta-data to properly interpret. These attributes imply that developing a sustained ocean observing system for climate will require partnerships between agencies and nations that share common interests. The investment requirements for continuity and long-duration measurements, coupled with regional- to global-scale spatial coverage, suggest that pooling and coordination of resources are the best strategy for sustaining ocean observations.

The Navy historically supported large-scale ocean measurement programs

---

[5]National Research Council. 2009. *Scientific Value of Arctic Sea Ice Imagery Derived Products,* The National Academies Press, Washington, D.C.

[6]See D.A. Rothrock, Y. Yu, and G.A. Maykut, 1999, "Thinning of the Arctic Sea-Ice Cover," *Geophysical Research Letters,* Vol. 26, No. 23, pp. 3469-3972; and D.A. Rothrock, D.B. Percival, and M. Wennahan, 2008, "The Decline in Arctic Sea-Ice Thickness: Separating the Spatial, Annual and Interannual Variability in a Quarter Century of Submarine Data," *Journal of Geophysical Research,* Vol. 113, C05003.

through Office of Naval Research (ONR) funding. For example, in the 1970s and 1980s, as part of the North Pacific Experiment (NORPAX), the Navy and NOAA jointly supported ship-of-opportunity and other measurements to document large-scale temperature anomalies in the North Pacific that were believed to affect North American weather and climate. However, over the past two decades, the Navy has greatly reduced its support of such large-scale ocean measurement efforts related to climate.

FINDING 6.1: The interconnectedness of the global ocean circulation, the involvement of processes spanning the full water column, the requirement to measure coast-to-coast across ocean basins, and the need for continuous long-term records represent daunting challenges to advancing understanding of climate variability and change. In view of these challenges, U.S. civilian agencies, in collaboration with international partners, have established a framework dating back to the early 1990s to advance the development of the Global Ocean Observing System (GOOS). Over the same time period, the U.S. Navy has withdrawn its support for large-scale ocean measurement programs; at present, it has little involvement in Global Ocean Observing System development. The Navy relies almost entirely on civilian agencies and their international partners for global-scale climate-related ocean measurement programs, which may fail to address specific Navy concerns.

RECOMMENDATION 6.1: The Office of Naval Research should reevaluate its long-standing decision to not support large-scale ocean measurement programs and instead participate directly in the large-scale sustained measurement programs that would support development of the Global Ocean Observing System.

FINDING 6.2: Open access to previously classified Navy data and to other Department of Defense assets through the MEDEA Program has enabled advances in climate change research that have benefited the scientific community studying climate change. A clear example of this benefit is the analysis of submarine upward looking sonar, which shows that sea ice has been thinning in response to climate change.

RECOMMENDATION 6.2: The Chief of Naval Research, the Oceanographer of the Navy, and the Commander, Naval Meteorology and Oceanography Command, should consider findings by the MEDEA Program (and take lessons from MEDEA actions within the intelligence community) to develop and support a Navy philosophy for providing access to previously classified information that can be used by the climate research community. Such actions would enhance the potential of these researchers to help the Navy better prepare for its mission in a future with a warmer climate.

Current Status of Climate Change Modeling

Climate modeling has rapidly evolved in recent years as a result of higher spatial resolution; better representation of physical processes; coupling of atmosphere, ocean, and land components; and the availability of a diverse array of observations. As the U.S. Navy considers the potential impact of climate change on its operations and national security, it is clear that evolutionary and transformational advances may be required to improve modeling and prediction of seasonal, decadal, and longer-term (century scale) climate. In particular, projections must provide sound estimates of the probability of the opening of the Arctic seas and high-risk events such as hurricanes, drought, and flooding. Sea-level rise and its consequences are also a primary model estimate of importance for operational planning beyond the next few decades.

Climate model projections based on external forcing of the climate system and predictions that, in addition, take into account initial conditions will be of great value for planning national security strategies in response to climate change and its impacts in the next few decades. Decadal prediction will be a new focus of the next Intergovernmental Panel on Climate Change (IPCC).[7,8] One of the challenges is determining how to properly initialize ocean and sea-ice models with observed climate conditions. The approach is a natural merger of ongoing efforts in seasonal-interannual and decadal forecasting.[9] NOAA's Climate Prediction Center (CPC) is a leading agency in the development of a robust Climate Forecast System and National Multi-Model Ensemble (NMME) system.[10] Involvement in these new efforts by the Navy at this planning stage could ensure that they are designed to meet the needs of the naval forces.

It is still quite challenging to assess climate model value or success because simulations from even the most advanced modeling systems have considerable spread and uncertainty. Carefully quantifying this uncertainty, especially at the regional scale, is necessary to evaluate the potential impacts of climate change. One effort is to create large ensembles of model simulations by varying uncertain

---

[7] Gerald A. Meehl, Lisa Goddard, James Murphy, Ronald J. Stouffer, George Boer, Gokhan Danabasoglu, Keith Dixon, Marco A. Giorgetta, Arthur M. Greene, Ed Hawkins, Gabriele Hegerl, David Karoly, Noel Keenlyside, Masahide Kimoto, Ben Kirtman, Antonio Navarra, Roger Pulwarty, Doug Smith, Detlef Stammer, and Timothy Stockdale. 2009. "Decadal Prediction: Can It Be Skillful?" *Bulletin of the American Meteorological Society,* pp. 1465-1485.

[8] James Hurrell, Gerald A. Meehl, David Bader, Thomas L. Delworth, Ben Kurtman, and Bruce Wielecki. 2009. "A Unified Approach to Climate System Prediction," *Bulletin of the American Meteorological Society,* pp. 1819-1832.

[9] Alberto Troccoli and T.N. Palmer. 2007. "Ensemble Decadal Predictions from Analyzed Initial Conditions," *Philosophical Transactions of the Royal Society A,* 365, No. 1857.

[10] University Corporation for Atmospheric Research. 2009. *Community Review of the NCEP Climate Prediction Center,* December. Available at http://www.ncep.noaa.gov/director/ucar_reports/CPC_Report_UCAR_Final.pdf. Accessed August 2, 2010.

physical parameters.[11] Progress is being made to improve climate models, but current modeling efforts suffer from insufficient resolution of features at various scales, including fronts, tropical-mid-latitude interactions, atmosphere-ocean exchanges, winds, precipitation, and salinity. Ocean components of climate models are relatively laminar (not turbulent) and do not capture energetic eddy flows at the sub-10 km scale.[12] New advances are also needed with respect to glacial ice. Current glacial models coupled to atmosphere or ocean models only account for superficial melting and accumulation rather than potential dynamic discharge by glacial flow, which may be critical for understanding future sea level.

The next generation of climate models must properly resolve carbon and other biogeochemical cycles to reach the goals of an Earth system modeling framework.[13] Uncertainties about carbon-feedback processes in the ocean and on land must be resolved to improve future predictions of climate change. The next IPCC assessment will include coupled modeling systems with interactive carbon cycles.[14] Refinements in ocean biochemistry modules in climate models are also required.

Expanding climate models to include these new processes with credible methods has resulted in a massive increase in the computing requirements, even at the standard resolutions of the IPCC *Fourth Assessment Report* (AR4). Increasing resolution will improve parameterization of clouds, ocean mixing, and ice sheets, but current high-performance computing resources are insufficient to resolve major issues associated with these processes. Also, the role of aerosols

---

[11]J.M. Murphy, D.M. Sexton, D.N. Barnett, G.S. Jones, M.J. Webb, M. Collins, and D.A. Stainforth. 2004. "Quantification of Modeling Uncertainties in a Large Ensemble of Climate Change Simulations," *Nature,* Vol. 430, pp. 768-772.

[12]David C. Bader, Curt Covey, William J. Gutowski, Isaac M. Held, Kenneth E. Kunkel, Ronald L. Miller, Robin T. Tokmakian, and Minghua H. Zhang. 2008. "Climate Models: An Assessment of Strengths and Limitations," synthesis and assessment product 3.1, U.S. Climate Change Science Program and the Subcommittee on Global Change Research Report. Available at http://www.climatescience.gov/Library/sap/sap3-1/final-report/default.htm. Accessed June 4, 2010.

[13]See National Research Council, 2007, *Earth Science and Applications from Space: National Imperatives for the Next Decade and Beyond,* The National Academies Press, Washington, D.C.; and National Research Council, 2008, *Earth Observations from Space: The First 50 Years of Scientific Achievements,* The National Academies Press, Washington, D.C.

[14]M.A. Shapiro, Jagadish Shukla, Gilbert Brunet, Carlos Nobre, Michael Béland, Randall Dole, Kevin Trenberth, Richard Anthes, Ghassem Asrar, Leonard Barrie, Philippe Bougeault, Guy Brasseur, David Burridge, Antonio Busalacchi, Jim Caughey, Deliang Chen, John Church, Takeshi Enomoto, Brian Hoskiins, Øystein Hov, Arlene Laing, Hervé Le Treut, Jochem Marotzke, Gordon McBean, Gerald Meehl, Martin Miller, Brian Mills, John Mitchell, Mitchell Moncrieff, Tetsuo Nakazawa, Haraldur Olafsson, Tim Palmer, David Parsons, David Rogers, Adrian Simmons, Alberto Troccoli, Zoltan Toth, Louis Uccellini, Christopher Velden, and John M. Wallace. 2010. "An Earth-System Prediction Initiative for the 21st Century: An International Interdisciplinary Initiative to Accelerate Advances in Knowledge, Prediction, Use and Value of Weather, Climate and Earth-System Information," National Center for Atmospheric Research, Boulder, Colo. Available at http://www.cgd.ucar.edu/.../ShapiroetalVisionDocument_FINALJan13_2010.pdf. Accessed June 4, 2010.

(liquid and solid particles in the atmosphere) is a major source of uncertainty in quantitative attribution and observational studies. Aerosols affect both radiation balance and cloud microphysical processes, yet they are very poorly represented in climate models.

For a meaningful translation of climate model information to regional and societal applications, current output is still on relatively coarse spatial scales. The well-known inconsistency between models' spatial resolution and scale of impact/decision making is challenging. And while computing resources will continue to improve spatial resolution and representation of physical processes, downscaling techniques (both dynamical and statistical) will likely be needed to overcome scale mismatches. Further, climate models should be more strongly coupled with decision support tools, models, and information systems that nonscientists and stakeholders use for decision making. The U.S. Navy is a good example of a stakeholder that has very specific needs in applications related to its infrastructure and operations, disease, civil instability, migration, water resources, and energy. A "holistic" modeling approach spans the climate, weather, and human dimension scale; it requires a seamless integration of chemistry, physics, climatology, meteorology, mathematics, and social and decision sciences.[15] An example of progress in this area is the University Corporation for Atmospheric Research (UCAR) Africa initiative on tropical health-climate-weather linkages.[16]

The U.S. Navy is not involved in coupled climate modeling or climate forecasting on any timescale. Most Navy modeling focuses on short tactical timescales and space scales for day-to-day fleet operations.[17] The Naval Research Laboratory (NRL) at Stennis Space Center, Mississippi, runs ocean models forced with prescribed atmospheric boundary conditions so that there are no feedbacks between the ocean and the atmosphere as in a truly coupled system, other than for 5-day sea-ice forecasting in the Arctic, done with the dated polar ice projection system. Model forecast systems under development at NRL are for the ocean only and are intended to infuse new technology into the Navy operational ocean forecasting at CNMOC; generally these model forecasts extend to 7 days in the future. Weather forecasting is carried out at FNMOC in support of fleet operations, but does not extend past 5-day lead times. In short, there is no capability for coupled ocean-atmosphere-land-cryosphere modeling in the Navy, and there are no programs focused on seasonal-to-decadal timescale predictions to support strategic decisions related to operations, platforms, and facilities. Because of the U.S. Navy's presence on the global oceans, its long-term global ocean/ice observations and

[15]Ibid.

[16]Information on the UCAR Africa initiative is available at http://www.africa.ucar.edu/index.html. Accessed June 4, 2010.

[17]These models include the Naval Research Laboratory's real-time Global Ocean Analysis and Modeling and 120-hour forecasts of ocean and sea ice made with the Polar Ice Prediction System and the Cox ocean model or the operational Arctic sea-ice charting done in collaboration between the National Ice Center and NOAA.

data collection, and its unique physical assets, the Navy can both benefit from and contribute strongly to a better understanding of the ocean component of climate science.[18]

Sea-Level Rise Modeling Needs

In the ranking of the top 20 cities in terms of population with projected exposure to coastal flooding in the 2070s, Miami ranks 6th and New York City 17th; the other 18 cities are all located in southeast Asia.[19] The total estimated 2070 population exposed to coastal flooding in the 10 most vulnerable cities tops 80,000,000 people. When smaller exposed coastal cities for these same nations are tallied, the total exposed population that could be in need of HA/DR assistance due to coastal storm damage is higher by two- to threefold (see Figure 6.1). The need for better models and understanding of coastal vulnerabilities thus has broader implications for naval forces than simply understanding the risks associated with naval coastal infrastructure.

One particularly good example of the potential of a vulnerability analysis to guide decisions relating to coastal exposure to storm surge in a future warmer world is that undertaken by Kelinosky et al.[20] The focus of their study was Hampton Roads, Virginia, at the nexus of the York and James Rivers, the Chesapeake Bay, and the Atlantic Ocean.

Researchers in the Hampton Roads study mapped physical exposure to storm-surge flooding for all categories of hurricane, for both present and future sea levels, using what would today be judged as conservative estimates for the latter. A total of 57 variables derived from the 2000 United States Census were used in a principal components analysis of social vulnerability. Maps of socioeconomic characteristics commonly associated with vulnerability to environmental

---

[18]For example, tracer-transport inversion is one of several methods for estimating greenhouse gas emissions, and it is based on atmospheric and/or oceanic measurements of the gases and mathematical models of air and water flow. Tracer-transport inversion estimates the net sum of anthropogenic and natural sources and sinks. The use of this method has been hypothesized as being potentially useful in estimating greenhouse gas emissions to support anticipated future climate treaty obligations. If such a scenario were to develop, this committee speculates that the U.S. Navy's instrument deployment and ocean data collection capabilities might also play a potential supporting role. See National Research Council, 2010, *Verifying Greenhouse Gas Emissions: Methods to Support International Climate Agreements,* The National Academies Press, Washington, D.C.

[19]R.J. Nicholls, P.P. Wong, V.R. Burkett, J.O. Codignotto, J.E. Hay, R.F. McLean, S. Ragoonaden, and C.D. Woodroffe. 2007. "Coastal Systems and Low-Lying Areas," *Climate Change 2007: Impacts, Adaptation and Vulnerability,* contribution of Working Group II to the *Fourth Assessment Report of the Intergovernmental Panel on Climate Change,* M.L. Parry, O.F. Canziani, J.P. Palutikof, P.J. van der Linden, and C.E. Hanson (eds.), Cambridge University Press, Cambridge, United Kingdom, pp. 315-356.

[20]Lisa R. Kelinosky, Brent Yarnal, and Ann Fisher. 2007. "Vulnerability of Hampton Roads, Virginia, to Storm-Surge Flooding and Sea-Level Rise," *Natural Hazards,* Vol. 40, No. 1, pp. 43-70.

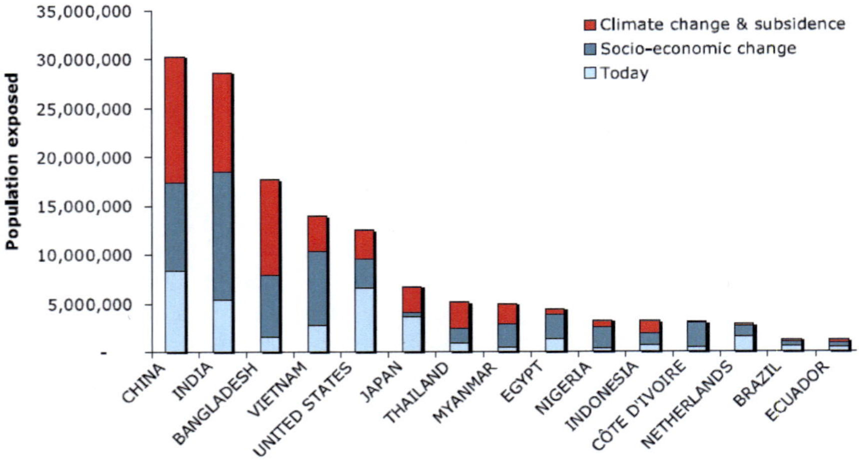

FIGURE 6.1 Top 15 countries by population exposed today and in the 2070s to coastal flooding, showing the influence of future climate change and socioeconomic change. SOURCE: R.J. Nicholls, S. Hanson, C. Herweijer, N. Patmore, S. Hallegatte, Jan Corfee-Morlot, Jean Chateau, and R. Muir-Wood. 2007. *Ranking of the World's Cities Most Exposed to Coastal Flooding Today and in the Future,* Organisation for Economic Co-operation and Development (OECD), Paris. Courtesy of OECD, 2007, "Ranking Port Cities with High Exposure and Vulnerability to Climate Extremes: Exposures Estimates," Environment Working Paper No. 1, available at http://www.oecd.org/env/workingpapers.

hazards are compared to the flood-risk exposure zones to identify the locations of vulnerable subpopulations. Scenarios that address uncertainties regarding future population growth and distribution are also developed to provide guidance that could help to diminish the vulnerability of future inhabitants of any metropolitan region to storm-surge flooding.

A summary statement in the U.S. Climate Change Research Program (2009) report on sea-level rise clearly describes the urgent need for new work on this topic: "The prospect of accelerated sea-level rise and increased vulnerability in coastal regions underscores the immediate need for improving our scientific understanding of and ability to predict the effects of sea-level rise on natural systems and society. These actions, combined with development of decision support tools for taking adaptive actions and an effective public education program, can lessen the economic and environmental impacts of sea-level rise."[21]

---

[21] James G. Titus, Eric K. Anderson, Donald R. Cahoon, Stephen Gill, Robert E. Thieler, and Jeffress S. Williams. 2009. *Coastal Sensitivity to Sea-Level Rise: A Focus on the Mid-Atlantic Region,* U.S.

THE SPECIAL CASE FOR UNDERSTANDING CHANGES IN THE ARCTIC

The retreat of Arctic sea ice in summer is fundamentally altering the naval forces' mission by allowing increasing access to the harsh and highly variable Arctic environment. As stated earlier in this report, the Arctic Ocean is in many ways the most poorly observed of the world's oceans: there are deficiencies in bathymetric charts, sparse knowledge of sea-ice thickness, infrequent measurements of ocean salinity and temperature, and so on. Current efforts to establish the first comprehensive, sustained in situ observing system in the Arctic are reviewed in this section, as is the state of climate modeling and seasonal forecasts of sea ice and Arctic climate that could prove valuable for planning Arctic operations.

The decline of the yearly minimum sea-ice cover in September is more than 10 percent per decade during the satellite era (since 1979; see Figure 2.3 in the operations section), and the decline appears to be accelerating. According to submarine sonar estimates of draft, the thickness of ice decreased by more than a meter from 1980 to 2000.[22] The Canadian archipelago has never allowed ice-free passage in the historical record until two summers in this decade—a prediction that was made at the 2001 symposium on "Naval Operations in an Ice-Free Arctic."[23] Observations and models indicate the Arctic ice cover is losing its multiyear ice and transitioning to a situation more like the Antarctic, which is mostly first-year ice covered with very little sea ice at the end of the melt season.[24,25]

Natural variability in the sea-ice extent is large in summer, so a given year can be far above or below (by at least 15 percent) the long-term trend. Explanations after the fact for the record minimum in 2007 are many-faceted,[26] which is evidence that a complete understanding of the mechanisms of sea-ice variability remains elusive. Predicting such fluctuations would be valuable, and a nascent effort known as the Sea Ice Outlook Project summarizes the community effort to produce Arctic-wide and regional forecasts 2 to 4 months in advance (http://www.

Climate Change Science Program and the Subcommittee on Global Change Research, U.S. Environmental Protection Agency, Washington, D.C., p. ix.

[22]D.A. Rothrock, D.B. Percival, and M. Wensnahan. 2008. "The Decline in Arctic Sea-Ice Thickness: Separating the Spatial, Annual and Interannual Variability in a Quarter Century of Submarine Data," *Journal of Geophysical Research*, Vol. 113, C05003.

[23] See Office of Naval Research, 2001, *Naval Operations in an Ice-Free Arctic Symposium: Final Report,* Arlington, Va., April.

[24]Josefino C. Comiso. 2002. "Warming Trends in the Arctic from Clear Sky Satellite Observations," *Journal of Climate,* Vol. 16, pp. 3498-3510.

[25]J.A. Maslanik, C. Fowler, J. Stroeve, S. Drobot, J. Zwally, D. Yi, W. Emery. 2007. "A Younger, Thinner Arctic Ice Cover: Increased Potential for Rapid, Extensive Sea-Ice Loss," *Geophysical Research Letters,* Vol. 34, L24501.

[26] Eric T. DeWeaver. 2008. "Arctic Sea Ice Decline: Introduction," *Arctic Sea Ice Decline: Observations, Projections, Mechanisms and Implications,* E.T. DeWeaver, C.M. Bitz, and B. Tremblay (eds.), pp. 1-6.

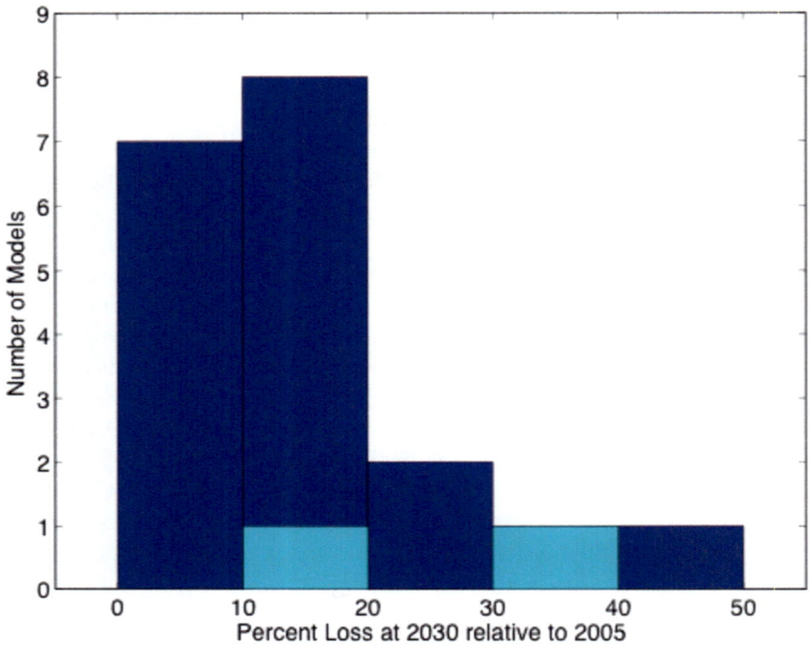

FIGURE 6.2 Percent loss at 2030 relative to 2005 (from 10-year means centered on these years) in models used for the *Fourth Assessment Report of the Intergovernmental Panel on Climate Change* (IPCC) from the Special Report on Emissions Scenarios (SRES) A1B (a balanced emphasis on all energy sources) scenario. Data were downloaded from the Coupled Model Intercomparison Project Phase 3 website. The models highlighted in pale blue are the only two models that agree with the observed mean and trend during the satellite era. SOURCE: Data adapted from the Coupled Model Intercomparison Project Phase 3 website at http://www-pcmdi.llnl.gov/ipcc/about_ipcc.php. Accessed April 8, 2011.

arcus.org/search/seaiceoutlook/). The effort began in 2008, motivated in part by the failure to anticipate the record low in 2007.

Climate models uniformly project continued sea-ice reduction in the 21st century, with most having a greater rate of thinning than reduction in extent.[27] However, many models correlate poorly with the observed sea-ice cover in the 20th century.[28] The spread in future projections is very broad—as is illustrated in Figure 6.2 with a histogram of the percent loss of Arctic September sea ice in

[27]Kyle Armour, Cecilia M. Bitz, LuAnn Thompson, Elizabeth H. Hunke, submitted, "Controls on Arctic Sea Ice First-Year and Multi-Year Ice Survivability." American Geophysical Union, Fall Meeting, 2009.

[28]Juliette Stroeve, Marika M. Holland, Walt Meier, Ted Scambos, and Mark Serreze. 2007. "Arctic Sea Ice Decline: Faster Than Forecast," *Geophysical Research Letters,* Vol. 34, L09501.

2030 relative to 2005. Figure 6.2 illustrates two key issues for Arctic operations. First, the sea-ice projections from climate models today are so broad that clearly a risk management approach is needed. Second, about one-quarter of the models project faster decline in the next 20 years than has been observed during the satellite era. The two models that agree with observations during the satellite era have above-average decline among models in the future. These same two models have ice-free conditions (i.e., the area falls below 1 million square kilometers) in September by roughly the years 2040 to 2060.

The spotlight on sea-ice projections from the last IPCC report (AR4) is likely to cause a step-change improvement in sea-ice modeling. The most profitable avenue of improvement is likely to be realized from improving the sea-ice climatology through tuning the model, as the climatology has been shown to have a substantial bearing on the subsequent trend.[29,30] There is likely to be substantial value from improving processes in the sea-ice component as well.[31]

Stimulated by the magnitude of Arctic climate change since the mid-1990s, the research community has been arguing for the need for a large-scale, sustained Arctic observing system. These efforts have culminated in two initiatives: (1) the U.S.-led Study of Environmental Change (SEARCH), and (2) the European Union-led Developing Arctic Modeling and Observing Capabilities for Long-Term Environmental Studies (DAMOCLES). Both programs were timed to coordinate major efforts during the International Polar Year of 2007-2008 and are at present working to leave in place long-term observing systems. The U.S. effort is spearheaded by the NSF and has led to development of the Arctic Observing Network (AON). The AON primarily provides data from NSF-sponsored investigators. The Arctic Council has organized a project known as Sustaining Arctic Observing Networks, or SAON, which offers to help coordinate sustained observations and to serve as a data portal (www.arcticobserving.org). The naval forces would benefit from being involved with these planning efforts.

FINDING 6.3: The Navy has billions of dollars in assets exposed to the threats of climate change, and it must make strategic decisions in the face of considerable uncertainty about the pace, magnitude, and regional manifestations of climate change. Yet Navy research at present has no capability for modeling the coupled ocean-atmosphere-land-cryosphere system and how it will respond to greenhouse gas forcing. The Navy also has no programs in seasonal-to-decadal timescale

---

[29]Cecilia M. Bitz. 2008. "Some Aspects of Uncertainty in Predicting Sea Ice Thinning," *Arctic Sea Ice Decline: Observations, Projections, Mechanisms and Implications,* E.T. DeWeaver, C.M. Bitz, and B. Tremblay (eds.), pp. 63-76.

[30]Julien Boe, Alex Hall, and Xin Qu. 2009. "September Sea-Ice Cover in the Arctic Ocean Projected to Vanish by 2100," *Nature Geoscience,* Vol. 2, pp. 341-343.

[31]Cecilia M. Bitz, J.K. Ridley, M.M. Holland, and H. Cattle. 2010. "20th and 21st Century Arctic Climate in Global Climate Models," in press in *Arctic Climate Change—The ACSYS Decade and Beyond,* P. Lemke (ed.).

climate forecasting to help guide long-range strategic planning for operations, platforms, and facilities; it relies almost entirely on civilian agencies and international assessments to inform its policies and practices related to climate change.

RECOMMENDATION 6.3: The Assistant Secretary of the Navy for Research, Development, and Acquisition (ASN RDA) should examine the U.S. Navy's overall research and development capabilities vis-à-vis climate studies, especially with respect to coupled models and climate forecasting on seasonal-to-decadal timescales. The ASN RDA should give special emphasis to regional aspects of sea-level rise, and sea-ice concentration and extent, because of their relevance to coastal infrastructure and operational needs. The Department of the Navy should also become actively engaged in the development of an Arctic Observing System, specifically with respect to development and deployment of in situ and remote sensing systems (i.e., gliders, buoys, and satellites) as well as icebreakers in support of research.

Appendixes

A

Terms of Reference

At the request of the Chief of Naval Operations, the Naval Studies Board of the National Research Council will establish a committee to study the national security implications of climate change for U.S. naval forces (i.e., the U.S. Navy, Marine Corps, and Coast Guard). Based on the Intergovernmental Panel on Climate Change assessments and other subsequent relevant literature reviewed by the committee, the study will:

1. Examine the potential impact on U.S. future naval operations and capabilities as a result of climate change (e.g., how will U.S. future naval operations be impacted and what capabilities will be needed for U.S. future naval forces as a result of climate change? This includes an assessment of the U.S. Coast Guard and Marine Corps, and where the U.S. Navy might be required to supplement or augment their capabilities).
2. Assess the robustness of the Department of Defense's infrastructure for supporting U.S. future naval operations and capabilities in the context of potential climate change impacts (e.g., are there any U.S. military installations and/or forward-deployed bases providing support to U.S. naval forces that are potentially vulnerable as a result of climate change?).
3. Determine the potential impact climate change will have on allied force operations and capabilities (e.g., are there any allies who may need U.S. naval force support as a result of climate change? Conversely, which allied force operations and capabilities may U.S. naval forces wish to leverage as a result of climate change?).
4. Examine the potential impact on U.S. future naval antisubmarine warfare operations and capabilities in the world's oceans as a result of climate change;

specifically, the technical underpinnings for projecting U.S. undersea dominance in light of the changing physical properties of the oceans.

This 15-month study will produce two reports: (1) a letter report following the third full committee meeting that summarizes the immediate challenges for U.S. naval forces in addressing each of the four above areas, as well as recommends approaches to address these challenges; (2) a comprehensive report that addresses in greater depth the full terms of reference.

B

Acronyms and Abbreviations

ACB	Amphibious Construction Battalion
ACC	America's Climate Choices
ANZUS	Australia, New Zealand, United States Security Treaty
AON	Arctic Observing Network
AR4	*Fourth Assessment Report* (of IPCC)
ARG	Amphibious Ready Group
ASN RDA	Assistant Secretary of the Navy for Research, Development, and Acquisition
ASW	antisubmarine warfare
C4ISR	command, control, communications, computers, intelligence, surveillance, and reconnaissance
CCSM3	Community Climate System Model version 3
CCSP	Climate Change Science Program (U.S.)
CDMA	Code Division Multiple Access
CIESIN	Center for International Earth Science Information Network (Columbia University)
CNA	Center for Naval Analyses
CNAS	Center for a New American Security
CNMOC	Commander, Naval Meteorology and Oceanography Command
CNO	Chief of Naval Operations
COCOM	Combatant Commander
COMSUBLANT	Commander, Submarine Force, Atlantic Fleet

CPC	Climate Prediction Center
CRMS	Center for Migration and Refugee Studies
CRS	Congressional Research Service
CVI	Coastal Vulnerability Index
DAMOCLES	Developing Arctic Modeling and Observing Capabilities for Long-Term Environmental Studies
DESDYNI	deformation, ecosystem structure, and dynamics of ice
DMSP	Defense Meteorological Satellite Program
DOD	Department of Defense
DOE	Department of Energy
DoN	Department of the Navy
DSCA	defense support of civil authorities
EAF	expeditionary airfield
EEZ	exclusive economic zone
EHF	extremely high frequency
ENSO	El Niño Southern Oscillation
ESA	European Space Agency
EUCOM	U.S. European Command
FAO	foreign area officer
FLTSAT	Fleet Satellite
FNMOC	Fleet Numerical Meteorology and Oceanography Center
FSU	former Soviet Union
GBS	Global Broadcast Service
GCOS	Global Climate Observing System
GDP	gross domestic product
GEO	geosynchronous Earth orbit
GERG	Geochemical and Environmental Research Group
GFCS	Global Framework for Climate Services
GLONASS	Global Orbiting Navigation Satellite System
GNSS	Global Navigation Satellite System
GOOS	Global Ocean Observing System
GPS	Global Positioning System
HA/DR	humanitarian assistance/disaster relief
HDOP	horizontal dilution of precision
HF	high frequency
HSPD	Homeland Security Presidential Directive
ICSU	International Council for Science

iGPS	High Integrity GPS Augmentation Demonstration Program (Naval Research Laboratory)
INMARSAT	International Maritime Satellite
INS	inertial navigation system
IOC	International Oceanographic Commission
IPCC	Intergovernmental Panel on Climate Change
JCS	Joint Chiefs of Staff
JHSV	joint high-speed vessel
JHU/APL	Johns Hopkins University Applied Physics Laboratory
LCS	littoral combat ship
LEO	low Earth orbit
LHA	amphibious assault ship (general purpose)
LHD	amphibious assault ship (multipurpose)
LIDAR	light detection and ranging
LPD	amphibious transport dock (ship)
MDA	maritime domain awareness
MEB	Marine Expeditionary Brigade
MEDEA	Measurements of Earth Data for Environmental Analysis
MEO	medium elliptical orbit
MEU	Marine Expeditionary Unit
MIT	Massachusetts Institute of Technology
MLP	Mobile Landing Platform
MODIS	Moderate Resolution Imaging Spectrometer
MPF	Maritime Prepositioning Force
MSC	Military Sealift Command
MUOS	Mobile User Objective System
NASA	National Aeronautics and Space Administration
NATO	North Atlantic Treaty Organization
NAVFESC	Naval Facilities Engineering Service Center
NAVOCEANO	Naval Oceanographic Office
NAVSEA	Naval Sea Systems Command
NEO	noncombatant evacuation order
NFH	Navy fleet hospital
NGO	nongovernmental organization
NIC	National Ice Center; National Intelligence Council
NMCB	Naval Mobile Construction Battalion
NMME	National Multi-Model Ensemble
NOAA	National Oceanic and Atmospheric Administration
NOC10	Naval Operations Concept 2010

NORPAX	North Pacific Experiment
NORTHCOM	U.S. Northern Command
NRC	National Research Council
NRL	Naval Research Laboratory
NSB	Naval Studies Board
NSE	Naval Support Element
NSF	National Science Foundation
NSIDC	National Snow and Ice Data Center
NSPD	National Security Presidential Directive
OCS	outer continental shelf
OECD	Office of Economic Cooperation and Development
ONR	Office of Naval Research
OSTP	Office of Science and Technology Policy
PACOM	U.S. Pacific Command
PEO-Ships	Program Executive Office for Ships
POM	Program Objective Memorandum
QDR	Quadrennial Defense Review
R&D	research and development
RF	radio frequency
RSR	refracted-surface reflected
SACEUR	Supreme Allied Commander Europe (NATO)
SAON	Sustaining Arctic Observing Networks
SAP 4.1	Synthesis and Assessment Product 4.1
SAR	synthetic aperture radar; search and rescue
SATCOM	satellite communications
SCICEX	Scientific Ice Exercise Program
SEARCH	Study of Environmental Arctic Change
SERDP	Strategic Environmental Research and Development Program
SOFAR	sound fixing and ranging
SPAWAR	Space and Naval Warfare Systems Command
SRES	Special Report on Emissions Scenarios
SSN	nuclear-powered attack submarine
SST	sea surface temperature
SUBICEX	submarine ice exercise
SUBPAC	Submarine Force, Pacific Fleet

TEC	total electron count
TFCC	Task Force Climate Change (Navy)
TOPEX	Ocean Topography Experiment
TOR	terms of reference
UAV	unmanned aerial vehicle
UCAR	University Corporation for Atmospheric Research
UCP	Unified Command Plan
UHF	ultrahigh frequency
UFO	UHF Follow-on (Satellite)
UN	United Nations
USCG	United States Coast Guard
USCGC	United States Coast Guard Cutter
USGS	U.S. Geological Survey
USMC	United States Marine Corps
UNCLOS	United Nations Convention on the Law of the Sea
USACE	U.S. Army Corps of Engineers
USN	United States Navy
VDOP	vertical dilution of precision
VHF	very high frequency
VRIM	Vulnerability-Resilience Indicators Model
WAAS	wide area augmentation system
WCRP	World Climate Research Program
WGS	Wideband Global Satcom
WHO	World Health Organization
WMO	World Meteorological Organization
WRS	wide area ground reference station
XBT	expendable bathythermograph
XCTD	expendable conductivity temperature and depth

C

Biographies of Committee Members and Staff

Frank L. Bowman (NAE), *Co-Chair*, is president of Strategic Decisions, LLC, a private consulting firm that provides governments and organizations with management training and strategic planning expertise. He retired from the U.S. Navy in December 2004 as a four-star admiral. He was director of the Naval Nuclear Propulsion Program and was concurrently deputy administrator—naval reactors in the National Nuclear Security Administration at the Department of Energy (DOE). In these dual positions, he was responsible for the operations of 103 reactors onboard the U.S. Navy's aircraft carriers and submarines, four training sites, and two DOE laboratories. His prior naval assignments include the chief of naval personnel and director of political-military affairs on the Joint Staff. Following his Navy career, he was president and chief executive officer of the Nuclear Energy Institute, the nuclear energy industry's policy organization. A graduate of Duke University, Admiral Bowman completed a dual master's program in nuclear engineering and naval architecture/marine engineering at the Massachusetts Institute of Technology (MIT) and was elected to the Society of Sigma Xi. He is a recipient of the Robert S. Landauer Memorial Lecture Award for distinguished contributions to the field of radiological physics and radiation health protection. He was awarded the honorary degree of Doctor of Humane Letters from Duke University. In 2006, Admiral Bowman was knighted by the Queen of England as Knight Commander of the Most Excellent Order of the British Empire in recognition of his commitment in support of the Royal Navy submarine program. He also has served on numerous advisory panels, including the MIT Nuclear Engineering Visiting Committee and as an advisor to the Pennsylvania State University's Nuclear Engineering Department. His National Research Council service includes being a member of the National Academy of Engineering's (NAE's) Forum on

Diversity in the Engineering Workforce (1999-2004). More recently, he served as a member of the military advisory board for the 2008 Center for Naval Analyses Corporation report entitled *National Security and the Threat of Climate Change*.

Antonio J. Busalacchi, Jr., *Co-Chair*, is director of the Earth System Science Interdisciplinary Center and Professor in the Department of Atmospheric and Oceanic Science at the University of Maryland, College Park. His research interests include tropical ocean circulation and its role in the coupled climate system, and climate variability and predictability. Dr. Busalacchi has been involved in the activities of the World Climate Research Program (WCRP) for many years and currently is chair of the Joint Scientific Committee that oversees the WCRP. He previously was co-chair of the scientific steering group for its subprogram on Climate Variability and Predictability. Dr. Busalacchi has participated extensively in National Research Council (NRC) activities, including as chair of the Climate Research Committee and the Committee on a Strategy to Mitigate the Impact of Sensor Descopes and Demanifests on the National Polar-orbiting Operational Environmental Satellite System and Geostationary Operational Environmental Satellite Spacecraft, and as a member of the Committee on Earth Studies, the Panel on the Tropical Ocean Global Atmosphere Program, and the Panel on Ocean Atmosphere Observations Supporting Short-Term Climate Predictions. Dr. Busalacchi currently serves as chair of the NRC Board on Atmospheric Sciences and Climate and as a member of the NRC Committee on America's Climate Choices: Panel on Advancing the Science of Climate Change.

Arthur B. Baggeroer (NAE) is Ford Professor of Engineering and the Secretary of the Navy/Chief of Naval Operations Chair for Ocean Science in the Departments of Ocean and Electrical Engineering at MIT. His areas of expertise include advanced signal processing methods applied to sonar, ocean acoustics, and geophysics. Dr. Baggeroer has served as director of the MIT–Woods Hole Joint Program in Oceanography and Oceanographic Engineering. During sabbatical leaves he has also served as a consultant to the Chief of Naval Research at the NATO Supreme Allied Commander Atlantic Center in La Spezia, Italy, and as a Green Scholar at the Scripps Institution of Oceanography. He is a fellow of the Institute of Electrical and Electronics Engineers and the Acoustical Society of America. Dr. Baggeroer recently served as co-chair of the NRC Committee on Distributed Remote Sensing for Naval Undersea Warfare. He is a former member of the NRC's Naval Studies Board and Ocean Studies Board.

Cecilia M. Bitz is an assistant professor in the Atmospheric Sciences Department at the University of Washington. Her research interests include climate dynamics, climate change, paleoclimate, the role of sea ice in the climate system, Arctic/North Atlantic interactions, and sea-ice model development. The primary tools for her research are a variety of models, from simple reduced models to sophisticated climate system models. Dr. Bitz is a member of the advisory board for the Community Climate System Model sponsored by the National Science Foundation and the Department of Energy and of the steering committee for the NOAA Climate and Global Change postdoctoral program. She currently serves on the NRC Climate Research Committee and was a member of the U.S. International Polar Year Planning Committee.

Ronald Filadelfo is research team leader for the Environment and Energy Research Team at the Center for Naval Analyses (CNA), the Navy's federally funded research and development center. His early work at CNA was in the field of antisubmarine warfare, and since the mid-1990s, he has worked on Navy environmental issues. His current research focus is on the effects of military sonars on marine mammals. For the last 2 years, Dr. Filadelfo has examined the issue of climate change and its relationships to national security. Most recently, he served on the CNA study team for the 2007 report entitled *National Security and the Threat of Climate Change.*

Jeffrey M. Garrett retired from the U.S. Coast Guard with the rank of Rear Admiral after 31 years on active duty. He is currently an independent consultant. Admiral Garrett served in a variety of command, operational, and staff assignments. His shipboard assignments included polar icebreaking deployments throughout the eastern and western Arctic and in the Antarctic in the polar icebreakers *Burton Island* (WAGB 283) and *Polar Star* (WAGB 10), and as commanding officer (CO) of *Polar Sea* (WAGB 11). As the commissioning CO of *Healy* (WAGB 20), he brought the nation's newest polar icebreaker through delivery, shakedown operations, and ice trials in the eastern Arctic, and through the Northwest Passage to homeport in Seattle. Other Coast Guard service included duties at the Vessel Traffic Service in Valdez, Alaska, command of a Great Lakes icebreaker, and multiple programming and budgeting staff assignments at Coast Guard headquarters. As a flag officer, Admiral Garrett served as the Coast Guard's director of resources, responsible for the service's budget, long-range planning, and policy development, and as Commander of the 13th District in the Pacific Northwest. Since retirement in 2005, Admiral Garrett has served as a member of the NRC Committee on Assessment of U.S. Coast Guard Polar Icebreaker Roles and Future Needs (2006), as chairman of the Coast Guard's Polar Operations and Policy study (2008), and as a consultant for a variety of maritime- and polar-related projects.

Harry W. Jenkins, Jr., retired from the U.S. Marine Corps with the rank of Major General and is currently an independent consultant. His background includes naval operations, mine countermeasures, and Marine Corps intelligence operations—in particular, its mission use of command, control, communications, computers, intelligence, surveillance, and reconnaissance, or C4ISR, systems. He formerly served as director of business development and congressional liaison at ITT Industries—Defense, where he was responsible for activities in support of tactical communications systems and airborne electronic warfare among the Navy, Marine Corps, Coast Guard, and National Guard. During Operation Desert Storm, General Jenkins served as the Commanding General of the Fourth Marine Expeditionary Brigade. He most recently served on the NRC Committee on the "1,000-ship Navy"—A Distributed and Global Maritime Network.

Catherine M. Kelleher is a professor of public policy at the University of Maryland and senior faculty associate at Brown University's Watson Institute, where her research interests include cooperative European defense and security policies, NATO relations, and international security and arms control. Dr. Kelleher served in the Clinton Administration as personal representative of the Secretary of Defense in Europe and as deputy assistant secretary of defense for Russia, Ukraine, and Eurasia. She has served on numerous scientific boards and advisory committees, including as vice chair, co-vice chair, and member of the NRC Committee on International Security and Arms Control and as a member of the NRC Committee on the "1,000-ship Navy"—A Distributed Global and Maritime Network. She is a member of the NRC's Naval Studies Board.

Mahlon C. Kennicutt II is a professor of chemical oceanography at the Texas A&M University, where he earned his Ph.D. in oceanography in 1980. After 18 months of postdoctoral work in geosciences at the University of Tulsa, Dr. Kennicutt returned to Texas A&M in 1981 and was a founding member of the Geochemical and Environmental Research Group (GERG). At GERG he served in various positions and rose to be director for 6-1/2 years, ending in 2004. He was promoted to full professor with tenure in the Department of Oceanography in 2002. At Texas A&M University he has served as a principal investigator (PI), deputy program manager, and/or program manager on several large interdisciplinary programs. He has participated in or led over $22 million worth of research, contracts, and grants during his time at Texas A&M University. He has served as PI on National Science Foundation (NSF) grants from Marine Chemistry and the Office of Polar Programs. He has spent over 575 days at sea, deployed to Antarctica six times, participated in six submersible dives in various vessels including the Navy NR-1, and maintains a current project at McMurdo Station, Antarctica, for the U.S. Army and NSF. Dr. Kennicutt served as leader of an interdisciplinary research program entitled the Sustainable Coastal Margins Program (SCMP), a coalition of six colleges, nine academic departments, five centers, and two institutes at Texas A&M as well as five partners external to Texas A&M. He was director of sustainable development in the Office of the Vice President for Research from 2004 to 2008. Dr. Kennicutt is the U.S. delegate to the Scientific Committee on Antarctic Research (SCAR). Dr. Kennicutt was elected to a 4-year term as president of SCAR in 2008. Within SCAR he has served on committees and held various offices, including that of vice president for finance and scientific affairs, member of the Standing Committee to the Antarctic Treaty System, chair of the Delegates Committee on Scientific Affairs, and secretary of the SCAR Scientific Research Program Subglacial Antarctic Lake Environments. He has also served as ex officio member of the Polar Research Board since 1998 and has been a science advisor to the U.S. Department of State Antarctic Treaty Delegations since 2002. Dr. Kennicutt has served as a member of several National Research Council committees and acted as a report monitor and external reviewer of several NRC reports. He is also served on a committee of visitors reviewing NSF Office of Polar Programs logistical and science support efforts.

Ronald R. Luman is head of the National Security Analysis Department at the Johns Hopkins University Applied Physics Laboratory (JHU/APL). He has led a variety of systems engineering efforts in the areas of missile guidance systems, unmanned undersea vehicles, countermine warfare, ballistic missile defense, intelligence systems, and domestic infrastructure resiliency. Dr. Luman has served on studies for the National Research Council (as a member of the NRC Committee on the Role of Naval Forces in the Global War on Terror) and for the National Infrastructure Advisory Council, and he chairs the annual JHU Unrestricted Warfare Symposium as a means to foster stronger collaboration among the national security policy, analysis, and technology communities. He is also the acting director of strategic planning at JHU/APL and the systems engineering program chair at the JHU Whiting School of Engineering.

W. Berry Lyons is a Distinguished Professor of Mathematical and Physical Sciences at the Ohio State University School of Earth Sciences and former director of the Byrd Polar Research Center. Previously he was a faculty member at the University of New Hampshire, the University of Nevada, Reno (where he served as the director of the Hydrologic Sciences Graduate Program), and the University of Alabama, Tuscaloosa (where he was the Loper Chair of Environmental Geology). Dr. Lyons's research interests include environmental geochemistry of trace metals, such as mercury; the causes and rates of chemical weathering and landscape change; the dynamics of carbon in the terrestrial environment; the role of agriculture and urbanization on water resources; and the impact of climate change on polar ecosystems. Dr. Lyons is a fellow of the Geological Society of America, the American Association for the Advancement of Science, and the American Geophysical Union. He is a past member of the NRC's Polar Research Board and past chair of the NRC Committee on Designing an Arctic Observing Network.

James J. McCarthy is the Alexander Agassiz Professor of Biological Oceanography at Harvard University. He holds faculty appointments in the Department of Organismic and Evolutionary Biology and the Department of Earth and Planetary

Sciences. His research interests focus on the regulation of plankton productivity in the sea and the upper ocean nitrogen cycle, especially in mixing processes, monsoonal cycles, and the El Niño Southern Oscillation system. He participated in the early planning phases of the International Geosphere-Biosphere Programme and served as its chair for the first 6 years of the program. He was involved in the first Intergovernmental Panel on Climate Change (IPCC) assessment, coauthoring the concluding chapter of Working Group I. In the third IPCC assessment he cochaired Working Group II, whose task it was to assess impacts of and vulnerabilities to global climate change, with an intensified focus on adaptation. He is past president and chair of the board of the American Association for the Advancement of Science. Dr. McCarthy has served on numerous scientific advisory boards and committees, including the NRC Ecosystems Panel, the Committee on Global Change Research, and the Committee to Review the Global Ocean Observing System.

Michael J. McPhaden is a senior scientist at the National Oceanic and Atmospheric Administration's Pacific Marine Environmental Laboratory in Seattle, Washington. He is an affiliate professor in the School of Oceanography at the University of Washington, director of the TAO [Tropical Atmosphere Ocean] Project Office, and a senior fellow with both the Joint Institute for the Study of the Atmosphere and Ocean at the University of Washington and the Joint Institute for Marine and Atmospheric Research at the University of Hawaii. Dr. McPhaden's primary research and expertise is in the ocean's role in climate and ocean-atmospheric interactions. He has served on the NRC Panel on Near-Term Development of Operational Ocean Observations. Dr. McPhaden is currently president of the American Geophysical Union.

John H. Moxley III (IOM) is retired managing director of North American Health Care Division, Korn/Ferry International. His areas of expertise include government policy and federal government administration; military training, costs, and manpower issues; military medical issues; and chemical, biological, radiological, and nuclear detection. He has held a number of senior positions in academia, government, and commercial industry, including dean of the University of Maryland and the University of California (San Diego) Medical Schools, assistant secretary of defense for health affairs, and senior vice president at American Medical International. He has served on numerous scientific boards and advisory committees, including the American Hospital Association Board of Trustees, the California Medical Association, the American Medical Association, the National Fund for Medical Education, and the Henry M. Jackson Foundation for the Advancement of Military Medicine. He recently served as cochair of the NRC Committee on Manpower and Personnel Needs for a Transformed Naval Force and is a former member of the NRC's Board on Army Science and Technology as well as the NRC's Naval Studies Board.

David J. Nash (NAE) retired from the U.S. Navy with the rank of Rear Admiral and is president of Dave Nash and Associates, LLC, a project development firm serving businesses and governments around the world. Admiral Nash has more than 4 decades of experience in building, design, and program management for both the U.S. Navy and the private sector. His experience includes the management of the Navy's shore installations worldwide and the reconstruction of Iraq's infrastructure. Most recently, Admiral Nash served as director of the Iraq Program Management Office under the Coalition Provisional Authority and, later, as director of the Iraq Reconstruction Management Office under the U.S. State Department. Admiral Nash is the recipient of numerous awards, including the Society of American Military Engineers Golden Eagle Award, the Beavers Award for Heavy Engineering Construction, the ASCE John I. Parcel-Leif J. Sverdrup Award for Civil Engineering Management, and the CERF/IIEC Henry L. Michel Award for Industry Advancement of Research. He was elected to the National Academy of Engineering "for leadership in the reconstruction of devastated areas after conflicts and natural disasters." He currently serves on the NRC Board on Infrastructure and the Constructed Environment and recently served on the NRC's Committee on Toward Sustainable Critical Infrastructure Systems: Framing the Challenges Workshop.

Heidi C. Perry is the director of Algorithms and Software at the Charles Stark Draper Laboratory. This engineering directorate supports business across the laboratory's wide spectrum of work in strategic, space, tactical, special operations, biomedical, and geospatial systems. Her expertise includes guidance, navigation and control, Global Positioning System anti-jam and ground control, autonomous vehicle (air, land, sea, space) avionics, and real-time embedded mission-critical software. She also has experience in developing signal processing systems, decision systems, and command, control, and communications systems. Previously, Ms. Perry served as director for Draper Independent Research and Development. She also served as the Mission Systems Division leader, Software Engineering Division leader, and principal member technical staff at Draper. She served as a member of the NRC Committee on the "1,000-ship Navy"—A Distributed Global and Maritime Network and is a member of the NRC's Naval Studies Board.

J. Marshall Shepherd is an associate professor at the University of Georgia, Athens (UGA), where he conducts research, advising, and teaching in atmospheric sciences, climatology, water-cycle processes, and urban climate systems. He joined UGA's Department of Geography's Atmospheric Sciences Program in January 2006. Prior to joining the UGA faculty, Dr. Shepherd spent 12 years as a research meteorologist in the Earth-Sun Division at NASA-Goddard Space Flight Center. Dr. Shepherd also served as a member of the United Nations World Meterological Organization steering team on aerosols and precipitation, and was contributing author on the 2007 IPCC *Fourth Assessment Report*. In 2004

Dr. Shepherd received a Presidential Early Career Award for Scientists and Engineers at the White House. He was made a fellow of the American Meterological Society in 2009. Dr. Shepherd is currently a member of the NASA Precipitation Measurement Missions Science team and the NOAA Climate Working Group. He is editor of the *Journal of Applied Meteorology and Climatology* and the climatogy section of *Geography Compass*.

Charles F. Wald retired from the U.S. Air Force as a four-star general with more than 35 years of service and more than 3,600 flying hours and 430 combat hours as a command pilot. General Wald is director and senior advisor to aerospace and defense industry for Deloitte Services, LLP. In this role, he is responsible for providing senior leadership in strategy and relationships with defense contractors and Department of Defense program executives. He is a subject matter specialist in weapons procurement and deployment, counterterrorism, national energy, and international security policy. General Wald has received major military awards and decorations, including the Defense Distinguished Service Medal, the Defense Superior Service Medal, and the Distinguished Flying Cross. A graduate of North Dakota State University, he received a master's degree in international relations from Troy University. He also completed coursework at Harvard University and the National War College and has been awarded an Honorary Doctor of Law degree from North Dakota State University. General Wald served as a member of the military advisory board for the 2008 Center for Naval Analyses report entitled *National Security and the Threat of Climate Change*.

Dr. David Whelan (NAE) is vice president, deputy general manager, Advanced Systems and chief scientist, Integrated Defense Systems, at the Boeing Company. His areas of expertise include defense research, development, and enabling technologies such as autonomous vehicles and space-based moving target indicator radar systems. Prior to joining Boeing, he served as director of the Tactical Technology Office at the Defense Advanced Research Projects Agency. His high-technology development experience includes roles as a research physicist for the Lawrence Livermore National Laboratory and as a lead engineer at Northrop Grumman. Dr. Whelan has served on numerous scientific boards and advisory committees, including the Defense Science Board, the Air Force Scientific Advisory Board, and the NRC Committee on Research, Development and Acquisition Options for U.S. Special Operations Command. Dr. Whelan is vice chair of the NRC's Naval Studies Board. He also serves as a member of the NRC's America's Climate Choices: Panel on Advancing the Science of Climate Change.

Dr. Carl Wunsch (NAS) is the Cecil and Ida Green Professor of Physical Oceanography at MIT. His research interests include ocean observing technologies and the general circulation of the world's oceans and its implications for climate change. He has served on numerous scientific boards and advisory committees, including the NRC's Ocean Studies Board and the International Steering Group for the World Ocean Circulation Experiment. He is a foreign member of the Royal Society and a recipient of the American Geophysical Union's Macelwane Award and Maurice Ewing Medal, and the American Meteorological Society's Henry Stommel Medal.

Staff

Charles F. Draper is director of the National Research Council's Naval Studies Board. Before joining the NRC in 1997, he was the lead mechanical engineer at S.T. Research Corporation, where he provided technical and program management support for satellite Earth station and small satellite design. He received his Ph.D. in mechanical engineering from Vanderbilt University in 1995; his doctoral research was conducted at the Naval Research Laboratory (NRL), where he used an atomic-force microscope to measure the nanomechanical properties of thin-film materials. In parallel with his graduate student duties, Dr. Draper was a mechanical engineer with Geo-Centers, Inc., working on-site at NRL on the development of an underwater X-ray backscattering tomography system used for the nondestructive evaluation of U.S. Navy sonar domes on surface ships.

Billy M. Williams is a senior program officer with the National Research Council's Naval Studies Board. Prior to joining the NSB, he served in a similar capacity with the NRC's Board on Army Science and Technology, where he led projects associated with the U.S. Army's chemical demilitarization program. Mr. Williams retired as a global research and development director from the Dow Chemical Company in 2004 after 30 years of service. His career at Dow included directing analytical materials science and polymer fabrication laboratories in operations across the United States, Europe, and Asia. He also served as the company's director of external science and technology programs, with responsibility for developing and securing strategic technical partnerships with leading research universities, national laboratories, and federal agencies. Mr. Williams earned an M.S. degree in organic chemistry and has completed executive education programs at Indiana University and Harvard University.

D

Letter Report to the U.S. Navy

THE NATIONAL ACADEMIES
Advisers to the Nation on Science, Engineering, and Medicine

Naval Studies Board
500 Fifth Street, NW
Washington, DC 20001
Phone: 202 334 3523
Fax: 202 334 3695
E-mail: nsb@nas.edu
www.nationalacademies.org/nsb

April 12, 2010

ADM Gary Roughead, USN
Chief of Naval Operations
2000 Navy Pentagon
Washington, DC 20350-2000

Dear Admiral Roughead:

In your letter dated September 12, 2008, to National Academy of Sciences President Ralph Cicerone, you requested that the National Research Council's (NRC's) Naval Studies Board (NSB) conduct a study to assess the implications of climate change for the U.S. Naval Services. Accordingly, in August 2009, the NRC, under the auspices of its NSB, established the Committee on National Security Implications of Climate Change for U.S. Naval Forces.

The study's terms of reference, provided in Enclosure A of this letter report, were formulated by the Chief of Naval Operations (CNO) in consultation with the NSB chair and director. The terms of reference charge the committee to produce two reports over a 15-month period. The present report is the first of these, a letter report issued, as requested, following the third full committee meeting.

The terms of reference direct that this study be based on Intergovernmental Panel on Climate Change (IPCC) scenarios and other peer-reviewed assessments. Therefore, the committee did not address the science of climate change or challenge the scenarios on which the committee's findings and recommendations are based. In short, this letter report summarizes the immediate challenges for U.S. naval forces in addressing each of the four areas listed below and recommends approaches for addressing these challenges. The terms of reference direct that the committee in its two reports do the following:

> 1. Examine the potential impact on U.S. future naval operations and capabilities as a result of climate change . . . .
> 2. Assess the robustness of the Department of Defense's infrastructure for supporting U.S. future naval operations and capabilities in the context of potential climate change impacts . . . .

3. Determine the potential impact climate change will have on allied force operations and capabilities . . . .
4. Examine the potential impact on U.S. future naval antisubmarine warfare operations and capabilities in the world's oceans as a result of climate change; specifically, the technical underpinnings for projecting U.S. undersea dominance in light of the changing physical properties of the oceans.

This first report is very much an interim report that neither addresses in its entirety any one element of the terms of reference nor reaches final conclusions on any aspect of the potential implications of climate change. Instead, this report highlights issues brought to the committee's attention during its first three meetings that could have potential near-term impacts, impose a need for near-term awareness, or require near-term planning to ensure that longer-term naval capabilities are protected. The committee will continue its study during the coming months and expects to complete by mid-2010 its final report, which will address all of the elements in the study's terms of reference and explore many potential implications of climate change not covered in this letter report.

In its initial three meetings, the committee received a number of helpful briefings from commands across the U.S. Navy, the U.S. Marine Corps, and the U.S. Coast Guard,[1] as well as expert briefings from individuals working at a number of other government agencies, including the following: the National Ice Center, the National Intelligence Council, the National Oceanic and Atmospheric Administration (NOAA), the Department of Energy's Oak Ridge National Laboratory, the Office of Naval Research (ONR), and the United States Geological Survey (USGS). Additionally, the committee conducted data-gathering sessions on national security and climate-change-related issues with Columbia University's Center for International Earth Science Information Network (CIESIN); the Cooperative Institute for Research in Environmental Sciences (CIRES), sponsored by NOAA and the University of Colorado, Boulder; the United Kingdom's Ministry of Defence; the Pacific Institute for Studies in Development, Environment and Security; and the Woods Hole Oceanographic Institution. When combined with the collective knowledge of the committee, these briefings are considered to constitute a sufficient basis for development of the findings and recommendations offered by the committee in this report.

[1]In its first three meetings, the committee heard from the U.S. Navy, Marine Corps, and Coast Guard as follows (see Enclosure D for dates, places, and briefers): (1) the U.S. Navy (Navy Meteorology and Oceanography Command, Navy Task Force Climate Change, Navy Energy Coordination Office, Navy Task Force Energy, Office of the Deputy Chief of Naval Operations for Fleet Readiness and Logistics, Office of the Deputy Chief of Naval Operations for Information Plans and Strategy [N3/N5], Office of the Deputy Chief of Naval Operations for Integration of Capabilities and Resources [N81], Office of the Commander of the Naval Installations Command, and the Navy Quadrennial Defense Review [QDR] Integration Group); (2) the U.S. Marine Corps (the Office of the Facilities Branch Head and the Office of Environmental Management Section, Headquarters Command); and (3) the U.S. Coast Guard (Commandant of the Coast Guard; and the Office of Policy Integration, USCG Headquarters).

BACKGROUND ON NAVAL FORCES AND CLIMATE CHANGE

The leaders of the U.S. Navy, Coast Guard, and Marine Corps have recognized the potential impact of climate change on naval forces' missions and have positioned their organizations to make adaptive changes.[2] In this regard, the CNO has recognized the linkage between energy use and climate change by establishing two key task forces: the Navy Task Force Energy (charged with formulating a strategy and plans for reducing the Navy's reliance on fossil fuels—and thus for reducing carbon dioxide emissions, operational energy demands, and, potentially, energy costs);[3] and the Navy Task Force Climate Change (charged initially with developing a roadmap for Navy actions in the Arctic, and then with addressing longer-term Navy actions regarding global climate change policy, strategy, and plans).[4] This committee engaged with the Navy Task Force Energy and the Navy Task Force Climate Change and found that each is providing strong leadership on these issues across the Navy and the Department of Defense (DOD). Both task forces are well positioned in capability and credibility to continue strong contributions within the DOD.

It is also noteworthy that the U.S. Navy and its assets are recognized by the national technical community as a critical partner in advancing the understanding of climate science and related policy implications.[5] The committee strongly supports the continuation of dedicated efforts by the Navy to be engaged with and to help lead these advances, within the broader context of the DOD's responsibility to assess the effects of climate change on all DOD missions, capabilities, and facilities. The Navy brings significant historical experience and unique capabilities to this arena, and the committee views these assets and related advances as supporting the national security interests of the United States.

---

[2]A board of retired flag and general officers also recognized this impact and provided a broader perspective on the topic of national security and climate change. See Military Advisory Board, 2007, *National Security and the Threat of Climate Change*, CNA Corporation, Alexandria, Va.

[3]CAPT James L. Brown, USN, Director, Navy Energy Coordination Office, Office of the Deputy Chief of Naval Operations for Fleet Readiness and Logistics, "Navy Task Force Energy, Perspectives and Related Climate Change Initiatives," presentation to the committee, September 17, 2009, Washington, D.C.

[4]See Vice Chief of Naval Operations Memorandum 4000 Ser N09/9U103035, "Task Force Climate Change Charter," October 30, 2009.

[5]For example, both Navy and Coast Guard assets have been highly important in providing critical scientific data associated with both ice mass and ocean changes over extended periods. Also, the Medea Program, a project of the 1990s, has been highly valuable in providing sea-ice data from military and intelligence assets that would otherwise be unavailable in the civilian sector. See National Research Council, 2009, *Scientific Value of Arctic Sea Ice Data,* The National Academies Press, Washington, D.C. In another example, SCICEX (Scientific Ice Expeditions) was a 5-year program in which the Navy made available a Sturgeon-class, nuclear-powered attack submarine for unclassified science expeditions to the Arctic Ocean to gather ice-thickness measurements. Additional information on SCICEX is available at http://www.Ideo.columbia.edu/res/pi/SCICEX/. Accessed November, 24, 2009.

This committee has found in the assessments it has studied strong scientific evidence to support naval leadership's continuing to study and act on the implications of climate change and how they will affect naval missions and capabilities. Some areas of current scientific knowledge of climate change, however, lack the near-term specificity that the Navy may need for planning purposes. These areas include, for example, the rate of future sea-level rise, the exact timing of an ice-free Arctic, and reliable predictions of regional climate (given the current inability to project specific regional impacts). Considering it unlikely that very precise projections of climate change will be available over the next few years, this committee believes that the Navy should adopt a risk management approach to addressing these issues. Such an approach should include a range of contingency plans for the potential sudden onset of climate-induced severe-weather disasters.

FINDINGS AND RECOMMENDATIONS

The committee focused its initial assessments either on climate change issues that it believes will have the greatest effects on naval forces or on issues that may require immediate planning. The committee views global climate change as a long-term issue that will play out over the next several decades. However, because of the long lead times in developing and changing naval capabilities and because of the potential for global climate change to have a significant impact on future naval missions, near-term awareness, planning, and decisions are needed by the U.S. Navy, Marine Corps, and Coast Guard.

The following sections of this report, in which appropriate supporting data are provided, present the committee's findings and recommendations, at this stage of the study, under the following four key topics, which are embedded in the terms of reference: (1) naval capabilities and potential climate-change-related operational issues globally, together with the closely related matter of the role of allied partnerships in regard to such global operational issues; (2) climate change impacts on global naval installations; (3) naval capabilities and potential climate-change-related operational issues in the Arctic; and (4) climate-change-related technical issues impacting naval operations, particularly in the Arctic.

1. Naval Capabilities and Potential Climate-Change-Related Operational Issues Globally

Naval Forces Responses in Future Climate-Change-Related Events

There are numerous peer-reviewed projections of increasing global stress arising from the effects of climate change as well as from the combined effects of climate change and other global trends, such as projected global population growth.[6] Current climate-change-related projections found in recent peer-reviewed reports portray a range of effects.[7] In turn, these reports and scientific models suggest that these effects may lead to more severe or frequent droughts, floods, storms, and other events with negative consequences for food and water supplies, leading to possibly even greater stress on the expanded population.[8] Viewed from a national security standpoint, these changes would likely amplify stresses on weaker nations and generate geopolitical instability in already-vulnerable regions.[9] A range of military missions might be necessary as a result of such conditions, including the sorts of antipiracy and counterterrorism missions now being conducted off the waters of Somalia. However, the clearest implications are for humanitarian assistance and disaster relief (HA/DR) missions, which may increase in frequency, thereby straining military transportation resources and the supporting force structures. The U.S. Navy, as a forward-deployed force in position to reach targeted disaster-relief sites faster than other agencies can, will almost surely experience increased demand for U.S. naval forces' assistance if climate-related disasters

---

[6]In many regions of the world, the impact of climate change is likely to further exacerbate the preexisting stress on water supplies and the mounting pressures of population growth. For example, Columbia University's Center for International Earth Science Information Network has compiled information from IPCC assessments, the 2005 World Bank report *Natural Disaster Hotspots: A Global Risk Analysis*, and CIESIN's gridded world population data sets to present a projected geographic distribution of vulnerability in 2050. In presentations to the committee, CIESIN representatives reported that global population nearly doubled from 1968 through 2008, and that by 2048 it could grow another 40 percent, to more than 9 billion people, adding even greater stresses to water and food supplies. CIESIN also reports that population increases are fastest in areas most vulnerable to intense storms and flooding (e.g., coastal areas, islands, and river basins). The CIESIN analysis combines its population data sets with IPCC-projected climate-change-related vulnerabilities, economic data, and past disaster-related losses to identify areas at relative high risk from one or more hazards. See Robert S. Chen, Center for International Earth Science Information Network, Columbia University, "Human Dimensions of Climate Change," and Marc Levy, Center for International Earth Science Information Network, Columbia University, "Climate Change and U.S. National Security," presentations to the committee, November 19, 2009, Washington, D.C.

[7]For example, see Thomas R. Karl, Jerry M. Melillo, and Thomas C. Peterson, 2009, *Global Climate Change Impacts in the United States,* Cambridge University Press, New York.

[8]See Intergovernmental Panel on Climate Change, 2007, "Climate Change 2007: The Physical Science Basis," Working Group I contribution to the *Fourth Assessment Report of the Intergovernmental Panel on Climate Change* (S. Solomon, D. Qin, M. Manning, Z. Chen, M. Marquis, K.B Averyt, M. Tignor, and H.L. Miller [eds.]), Cambridge University Press, Cambridge, United Kingdom and New York. See also C.P. McMullen and J. Jabbour, 2009, *Climate Change Science Compendium,* United Nations Environment Programme, EarthPrint, Nairobi, Kenya.

[9]See June 25, 2008, House Permanent Select Committee on Intelligence and House Select Committee on Energy Independence and Global Warming: *Statement of the Record by Dr. Thomas Fingar, Deputy Director of National Intelligence for Analysis—National Intelligence Assessment on the National Security Implications of Global Climate Change to 2030*. Available at http:www.dni.gov/testimonies/20080625_testimony.pdf. Accessed November 24, 2009. See also Military Advisory Board, 2007, *National Security and the Threat of Climate Change*. CNA Corporation, Alexandria, Va.

increase.[10] The demand for Navy Construction Battalion capability in support of HA/DR operations is expected to increase in proportion to the operational tempo of U.S.-sponsored international HA/DR operations.[11] Likewise, the U.S. Marine Corps should expect that it will be called on as an expeditionary ground force to assist with extreme-weather-related HA/DR events in a changing climate and to help secure U.S. interests in sensitive regions.[12] However, the pace and extent of this increase are unknown.

Based on the current uncertainty regarding the pace and extent of this demand, the Navy should not at present fund changes to force structure for humanitarian assistance and disaster relief, but over time it should consider changes to the construction of future naval platforms of appropriate classes in order to accommodate HA/DR operations and potential increases in climate-change-related mass human migration. The benefit of the Navy's providing such HA/DR support was demonstrated in the 2004 tsunami relief effort in Indonesia and in the recent earthquake relief work in Pakistan and Haiti.[13] The U.S. Navy, Coast Guard, and Marine Corps need to consider the ramifications of this enhanced HA/DR mission and the ways to prepare for it, including regular reviews of advanced staging requirements. A possible near-term investment might be considered for increased Navy Construction Battalion capacity for such deployments. If such efforts are not

[10]A 2007 joint maritime strategy document for the U.S. Navy, Marine Corps, and Coast Guard calls out "humanitarian assistance and disaster response" as one of six capabilities that constitute the core of U.S. maritime power and that "reflect an increased emphasis on those activities that prevent war and build partnerships." See *Cooperative Strategy for 21st Century Seapower,* available at http://www.navy.mil/maritime/MaritimeStrategy.pdf. Accessed November 23, 2009. However, it is not the sole responsibility of the U.S. military to respond to national and international humanitarian and disaster-relief emergencies; many U.S. and international governmental and private agencies may be engaged in any given relief operation.

[11]For a review of U.S. Navy Construction Battalion operations, see U.S. Navy Seabees First Naval Construction Division, Strategic Plan 2008-2011, Norfolk, Va.

[12]For example, in the aftermath of Tropical Storm Ketsana striking the Philippines on September 25, 2009, the U.S. Navy and U.S. Marine Corps worked with the Philippine government (and in support of the U.S. Department of State and the U.S. Agency for International Development Office of Foreign Disaster Assistance) to rapidly provide critically needed supplies in support of disaster relief to help mitigate human suffering and prevent further loss of life. In this case, a team of approximately 100 personnel composed of Marines from 111 Marine Expeditionary Forces flew from Okinawa to the Philippines on September 29, 2009, to conduct humanitarian assistance assessments. On September 30, U.S. Navy ships USS *Denver*, USS *Tortuga*, and USS *Harpers Ferry*, with embarked Marines and sailors of the 31st Expeditionary Unit, set sail from Okinawa toward the Philippines. On October 1, the commanding general of the 3rd Marine Expeditionary Brigade flew from Okinawa to the Philippines to lead planning and humanitarian assistance efforts. See *U.S. Marine Corps News*. Available at http://www.okinawa.usmc.mil/public affairs/info/archive/news. Accessed November 23, 2009.

[13]See "U.S. Navy Relief Efforts After the Indian Ocean Tsunami," December 26, 2004, Department of the Navy—Navy Historical Center, available at http://www.history.navy.mil/faqs/faq130-4.htm; and "U.S. Navy Transports UAE Donation to Earthquake Victims in Pakistan," November 3, 2005, Department of the Navy news article, available at http://www.news.navy.mil/search/display.asp?story_id=20885. Accessed November 23, 2009.

planned already, U.S. naval forces could benefit from a full inventory and review of the lessons learned from recent HA/DR deployments, such as the U.S. Navy, Marine Corps, and Coast Guard deployment to provide HA/DR after the January 2010 earthquake in Haiti.

Allied Forces Responses in Future Climate-Change-Related Events

An issue closely related to U.S. naval capabilities and global response to projected climate change is the role of allied forces partnerships. The committee received briefings from the National Intelligence Council suggesting that, in addition to the security challenges discussed above, the impact of projected severe climate change on food or water supplies and on disease patterns in certain regions of the world may lead to large-scale regional population movements, resulting potentially in millions of what some have termed "climate refugees" fleeing environmental "hotspots."[14,15] These assessments suggest that if such large-scale movements were to develop, U.S. naval leadership should be prepared for the possibility that allied forces might be occupied by their own domestic and regional climate-change-related responses, or that allied forces might lack the appropriate response capacities to assist with international efforts. In such cases, the National Command Authority might require U.S. naval forces to act alone, without allied forces' assistance.

This committee has not yet fully explored the views, issues, and capabilities concerning climate change with respect to allied forces, nor has it conducted an assessment of projections involving associated climate-change-related geographic hotspots. The committee's early assessment of allied partnerships indicates that several countries, especially in Europe, are already assuming strong public military postures on climate change,[16] and those countries may be open to the

[14]MajGen Richard Engel, USAF (Ret.), Director, Climate Change and State Stability Program, National Intelligence Council, "National Intelligence Assessment on the National Security Implications of Global Climate Change to 2030," presentation to the committee, October 19, 2009, Washington, D.C.

[15]The term "climate refugee" in this report refers to persons who cross international borders because of drought, flooding, or other severe weather or extreme events related to climate change. Currently the term "climate refugee" or "environmental refugee" has no standing in international law; it is not defined with any entitled protection under the 1951 United Nations Refugee Convention or the 1967 UN Refugee Protocol, although there is a movement among many nongovernmental organizations to petition for this recognition. See Bonnier Docherty and Tyler Giannini, 2009, "Confronting a Rising Tide: A Proposal for a Convention on Climate Change Refugees," *Harvard Environmental Law Review*, Vol. 33, pp. 349-403.

[16]The committee received a briefing on the United Kingdom's Ministry of Defence climate-change-related policies and plans from the British Defence Staff of the United States British Embassy, and will pursue discussions with official representatives of other U.S. allies. Related to this, military experts from many nations are increasingly expressing concerns about the need for attention to climate-change-related national security. For example, see "Australian Military Warns of Climate Conflict," available at http://www.reuters.com/article/environmentNews/idUSTRE5060FU20090107. Accessed November 24, 2009. See also statements endorsed by military experts of the United Kingdom, the

establishment of cooperative partnerships for leveraging capabilities to meet potential global climate-change-related HA/DR needs. The committee plans to address more expansively in its final report issues related to allied partnerships, but it believes that early planning and engagement by U.S. naval leadership with allied partners to address climate change issues are called for.

Finding 1: Scenarios of global climate change from the Intergovernmental Panel on Climate Change project impacts on both developing and developed nations, and such impacts may be destabilizing in many parts of the world. These projected changes would affect U.S. national security and stress naval resources. In particular, naval forces might be required to carry out more frequent humanitarian assistance and disaster relief (HA/DR) missions. At the same time, U.S. naval forces would be expected to execute their ongoing national-security-related military missions and to position themselves for supporting missions in destabilized regions around the globe. Also, it is expected that the demand for U.S. Navy Construction Battalion capability would increase in proportion to the operational tempo of U.S. HA/DR operations.

Recommendation 1: Although the committee has not yet completed its full analysis of the implications of climate change for future Navy force structure, it is clear that the Chief of Naval Operations (CNO) needs increasingly to take such implications into account. The committee believes that the CNO should not in the near term specifically fund new force-structure capability to deal with the effects of climate change but should hedge against climate change impacts through planning for the modification of the existing force structure as the climate-change-related requirements become clearer. All of the U.S. naval forces (the U.S. Navy, Marine Corps, and Coast Guard) should begin to consider potential specific force-structure capabilities and training standards over the next 10 years for conducting missions arising from the effects of climate change. In particular, the Navy should review the current and projected Navy Construction Battalion capability and capacity in light of the potential acceleration of the current operational tempo as a result of climate change effects.

Netherlands, and India in the October 29, 2009, press release "Military Experts from Five Continents Warn of Impact of Climate Change on Security," Institute for Environmental Security, Washington, D.C.

2. Climate Change Impacts on Global Naval Installations

Global sea-level rise is projected to be a major impact of climate change.[17] Many naval coastal installations would be affected and would likely require adaptation.

Measurements show that the 20th-century rate of global average sea-level rise is about 2 millimeters (mm) per year, and the rate of sea-level rise since 1993 has been measured to be about 3 mm per year.[18] This acceleration is consistent with an increase in the contribution from the melting of mountain glaciers and ice sheets on Greenland. However, the U.S. Naval Services cannot assume that the recent rate of sea-level rise will remain steady for the rest of the 21st century. Prudent planning and routine reevaluation of the rate and pace of change are necessary. The projected increased intensity of storms and storm surges also contributes to the anticipated increased vulnerability of naval coastal installations.[19]

Sea-level rise is not uniform around the globe, and the potential coastal impact of regional sea-level rise is not linear with elevation. Because of shifts in ocean circulation and the redistribution of mass in the global ocean, regional sea-level changes can vary from the global mean. Indexes of coastal vulnerability should take into account factors such as coastal topography, the local rate of relative sea-level rise, subsidence, regional extreme-weather history, population density, local freshwater supply sources, and critical infrastructure—such as communications, transportation, and utilities.

The committee reviewed an assessment of U.S. military coastal installations at risk from coastal inundations caused by sea-level rise.[20] Many of the 31 U.S. military installations identified in the assessment as being at "very high risk" or at "high risk" are naval installations. These data provide a starting point for more in-depth evaluations. As directed by requirements for the DOD's Quadrennial Defense Review (QDR), a broader analysis of global military coastal installations is also being conducted at this time.[21] This broader QDR-driven assessment will

[17]See Thomas R. Karl, Jerry M. Melillo, and Thomas C. Peterson, 2009, *Global Climate Change Impacts in the United States,* Cambridge University Press, New York, pp. 25-47.

[18]Konrad Steffen, Director, Cooperative Institute for Research in Environmental Sciences, University of Colorado, "State of the Science for Sea-Level Rise Data," presentation to the committee, October 20, 2009, Washington, D.C.

[19]See Thomas R. Farl, Jerry M. Melillo, and Thomas C. Peterson, 2009, *Global Climate Change Impacts in the United States,* Cambridge University Press, New York, pp. 114, 149.

[20]For this particular assessment, the U.S. Geological Survey's Coastal Vulnerability Index (CVI) was used. This index estimates risks of impact related to sea-level rise using a set of factors such as rate of sea-level rise, tidal range, and coastal geomorphology. See MajGen Richard Engel, USAF (Ret.), Director, Climate Change and State Stability Program, National Intelligence Council, "National Intelligence Assessment on the National Security Implications of Global Climate Change to 2030," presentation to the committee, October 19, 2009, Washington, D.C.

[21]The 2010 Quadrennial Defense Review (QDR) is a legislatively mandated (USC 10, Sec. 118 [a]) review of the Department of Defense strategy and priorities. The review takes place every 4 years and will be provided to Congress in early 2010. For the first time, the 2010 QDR is explicitly

provide a foundation, but there is a clear need for a more detailed global analysis and an action plan to address the vulnerabilities of those coastal installations identified as being at very high risk and at high risk. The assumptions, decisions, and time lines for addressing these risks should be determined on a consistent basis across the DOD. The committee suggests that additional risk factors beyond the current indicators of sea-level rise, tidal range, and coastal geomorphology be included in future analyses: such additional risk factors as regional weather history and potential impacts on critical infrastructure, as outlined above, in addition to shifts in storm tracks, changes in ocean circulation, and the impact of groundwater drawdown and recharge on subsidence are critical. The committee believes that these analyses must explicitly address the broader issue of the potential for sea-level rise and more intense storm surges to impact critical military missions.

On the basis of presentations to the committee, there appear to be at least three separate Navy groups involved in the analysis of coastal-installation vulnerability issues for the Navy: the Space and Naval Warfare Systems Command (SPAWAR), the Naval Facilities Engineering Service Center (NAVFAC ESC),[22] and the Naval Installations Command.[23] Additionally, according to presentations provided to this committee, prior to the QDR request the U.S. Marine Corps began conducting an analysis of its U.S. coastal-installation vulnerabilities.[24] Also, the DOD's Strategic Environmental Research and Development Program (SERDP) has initiated climate-change-related military infrastructure studies.[25] This committee believes that to avoid duplication of effort and to ensure a more comprehensive and consistent assessment, a more coordinated vulnerability analysis is needed across the naval installations nationally and internationally and in conjunction with all Services.

Considering the current measurements for sea-level rise, a major resource investment is not anticipated to be required by the Navy in the near term (the next

---

asked to include climate-change trends in its address of the national strategic and security environment. See U.S. Department of Defense, "2010 QDR Terms of Reference Fact Sheet," April 27, 2009, Washington, D.C.

[22]See for example, Kathleen Paulson and Dallas Meggitt, 2008, *US Naval Facilities Engineering Service Center Environmental Program on Climate Change,* Naval Facilities Engineering Service Center, Port Hueneme, Calif.

[23]CAPT Brant Pickrell, USN, Deputy Director, Shore Readiness, Commander, Naval Installations Command, "Preliminary Climate Change Related Naval Base Assessments—A Status Report," presentation to the committee, October 19, 2009, Washington, D.C.

[24]Elmer W. Ransom, Environmental Management Section, Headquarters, U.S. Marine Corps, and Capt Anthony V. Ermovic, USMC Facilities Branch Head, Headquarters, U.S. Marine Corps, "Marine Corps Perspectives and Climate Change Initiatives," presentation to the committee, September 18, 2009, Washington, D.C.

[25]The Office of the Secretary of Defense's Strategic Environmental Research and Development Program currently sponsors several projects related to the assessment of the impact of global sea-level rise on military infrastructure. These projects are managed under SERDP's Sustainable Infrastructure Projects Program. Descriptive information on these projects (SI-1700, -1701, -1702, and -1703) is available at http:/www.serdp.org/Research/SI-Facilities-Management.cfm. Accessed November 23, 2009.

5 years), with the exception of those naval installations identified as being at very high risk. However, in the longer term (the next 20 to 30 years), investments will have to be made for the adaption of many naval coastal installations, and those investments may have implications for decisions being made today.[26] The committee will address this issue more fully in its final report, but it believes that the current preliminary naval coastal-installation vulnerability assessment underway in support of the QDR is a good starting point and reflects a prudent course of action. The committee has not reviewed detailed vulnerability data associated with Navy Base Diego Garcia and Navy Base Guam. However, based on publicly available coastal-elevation data, the committee believes that these two naval bases may require special short-term attention and, potentially adaptive measures.

Finding 2: U.S. Navy, Coast Guard, and Marine Corps coastal installations around the globe will become increasingly susceptible to projected climate events. Several assessments now being made of naval-installation vulnerabilities appear to be focused primarily on sea-level rise and coastal inundation only. According to these current assessments, some adaptive actions are indicated owing to already-identified vulnerabilities at specific naval installations. The preliminary review of climate-change-related base vulnerabilities across the DOD—currently underway as directed by the requirements for the Quadrennial Defense Review—does not include all of the important factors affecting coastal-installation vulnerabilities, but it does provide a baseline assessment across all Services and a starting point for more in-depth analysis.

Recommendation 2: The Commander, Naval Installations Command, and the Navy Director for Fleet Readiness and Logistics should work with their U.S. Coast Guard and Marine Corps counterparts—and in conjunction with other Services and the Office of the Secretary of Defense—to ensure that a coordinated analysis of naval mission vulnerability is undertaken in order to address naval-installation vulnerability to rising sea levels and higher storm surges. Such an approach is necessary to avoid duplication of effort and to ensure that a more uniform and comprehensive evaluation is undertaken. For Program Objective Memorandum (POM)-14 planning purposes, the CNO should prepare to invest in early-stage mitigation and adaptation for targeted low-elevation naval installations identified in current vulnerability assessments as being at very high risk from more intense storm surges and other climate impacts. Other risks for naval installations as a result of projected climate change require further analysis and planning at this time

[26]The Navy's needs, if any, would be reflected in Program Objective Memorandum (POM) submissions. The POM submission is a 5-year outlook on budget requirements. It starts with the year following the President's Budget, which is always 1 year ahead of the current year.

but no immediate direct additional substantial investment beyond current budget plans.

3. Naval Capabilities and Potential Climate-Change-Related Operational Issues in the Arctic

Projected global climate change may have its most immediate and obvious implications for maritime operations in the Arctic region. The Arctic provides dramatic evidence of recent trends in global climate change as demonstrated by the continued significant reduction in summer sea-ice cover in the Arctic Ocean and the rapid disappearance of older, thicker, multiyear ice.[27] A result of this change is greater summer marine access and longer seasons of potential navigation. The committee does not expect that there will be routine commercial shipping in the Arctic in the foreseeable future, but a notable increase in private tourism and exploration traffic through the region is already occurring.[28] In addition, the U.S. Geological Survey notes that significant natural resources (oil, natural gas, and nonfuel minerals) may become increasingly available for exploitation as ice melts.[29] The physical indicators at hand and the current model projections provide strong evidence that future requirements for U.S. maritime operations in the Arctic will increase over the next 30 years.[30] The committee offers the following initial observations regarding naval Arctic operations.

Key Arctic Operational Challenges

Operating in the Arctic is not, at present, a priority for the Navy, although an increase in Arctic presence and operations *is* a priority for the Coast Guard. Unclassified national intelligence assessments suggest a low likelihood of sig-

---

[27]On September 12, 2009, sea-ice extent reached a 2009 minimum of 5.1 million km^2. The summer minimum is the third-lowest recorded since 1979. While the 2009 minimum was an increase over that of the 2 previous years, it was still 1.6 million km^2 below the 1979-2000 average minimum. The March 2009 ice extent was 15.2 million km^2, the same as in 2008 and only 4 percent less than the 1979-2000 average of 15.8 million km^2. March is historically the month of maximum sea ice extent. See *Arctic Report Card: Update for 2009,* available at http://www.arctic.noaa.gov/reportcard/ArcticReportCard_full_report.pdf. Accessed November 24, 2009.

[28]Arctic Council. 2009. *Arctic Marine Shipping Assessment 2009 Report,* available at http://www.nrf.is/index.php/news/15-2009/60-arctic-marine-shipping-assessment-report-2009. Accessed November 24, 2009.

[29]See July 23, 2008, U.S. Geological Survey press release, "90 Billion Barrels of Oil and 1670 Trillion Cubic Feet of Natural Gas Assessed in the Arctic," available at http://www.usgs.gov/newsroom/article.asp?ID=1980&from=rss_home. Accessed November 23, 2009.

[30]An Arctic Roadmap has been prepared by the Navy Task Force Climate Change, and the committee was briefed on it. See CAPT(S) Timothy Gallaudet, USN, Deputy Director, Navy Task Force Climate Change, "Task Force Climate Change Update and Gaps and Projected Future Needs," presentation to the committee, October 19, 2009, Washington, D.C.

nificant conflict in the Arctic region in the foreseeable future.[31] However, the demand for Coast Guard missions is already increasing. The committee believes that it would be prudent for the Navy and the Coast Guard to establish a strong and consistently funded technology, environmental data-gathering, and modeling foundation that can support a surge in operations.[32] Both the Navy and the Coast Guard should create a foundation of operational experience in the Arctic so that the challenges are well understood by a significant portion of U.S. naval forces in the following three areas:

1. *Port/Airbase Facilities.* Arctic distances are vast, and U.S. naval support infrastructure capabilities are sparse at best in that region. The distance from the southern coast of Alaska (where there are permanent U.S. Coast Guard bases) to the northern coast (where there are limited temporary facilities during the summer months) is approximately 800 nautical miles. The ability of the U.S. Coast Guard to execute its mission responsibilities in the Arctic is marginal, although there have been summer Coast Guard operational exercise surges in the Arctic during the past 3 years. The major review of Coast Guard requirements now underway will better inform the need for protecting U.S. national security interests in the Arctic.[33] Continued Coast Guard summer operations will strengthen the U.S. national presence and capabilities in the Arctic.

2. *Command, Control, Communications, Computers, Intelligence, Surveillance, and Reconnaissance (C4ISR) Infrastructure.* The robust set of geosynchro-

[31]See National Intelligence Council, *Global Trends 2025,* November 2008, p. 53, available at http://www.dni.gov/nic/PDF_2025/2025_Global _Trends_Final_Report.pdf. Accessed November 24, 2009. This unclassified report states in part: "Although serious near-term tension could result in small-scale confrontation over contested claims, the Arctic is unlikely to spawn major armed conflict. Circumpolar states have other major ports on other bodies of water, so the Arctic does not pose any lifeblood blockade dangers. Additionally, these states share a common interest in regulating access to the Arctic by hostile powers, states of concern or dangerous non-state actors; and by their shared need for assistance from high-tech companies to exploit the Arctic's resources."

[32]As examples of past Navy Arctic engagement, the Office of Naval Research's Arctic research funding has dropped from about $30 million per year in the early 1990s to about $3 million per year currently. Also, the U.S. Navy's Cold War Arctic Infrastructure no longer exists. See Richard F. Pittenger and Robert B. Gagosian, 2003, "Global Warming Could Have a Chilling Effect on the Military," *Defense Horizons*, No. 33, p. 7, October.

[33]National Security Presidential Directive (NSPD) 66 of January 2009 establishes the policy of the United States with respect to the Arctic region and outlines national security and homeland defense interests in the region. In part, NSPD 66, Article III B 1, states that these interests include "such matters as missile defense and early warning; deployment of sea and air systems for strategic sealift, strategic deterrence, maritime presence, maritime security operations, and ensuring freedom of navigation and overflight." The implementation of NSPD 66 will require multiagency and full government participation. For its part, the U.S. Coast Guard has commissioned a study—the USCG High Latitude Study anticipated to be completed in mid-2010—to better define its needs to support this Arctic directive, as the committee learned in a discussion with ADM Thad Allen, Commandant, U.S. Coast Guard, November 20, 2009, Washington, D.C.

nous Earth orbit (GEO) satellites provides reliable communications for locations below 65°N or above 65°S. High-data-rate satellite communications are sparse over the polar regions. However, commercial low-rate service is available. Additionally, Global Positioning System (GPS) constellation coverage is not optimized for polar regions, and so its accuracy is reduced, but still provides adequate performance for surface navigation. The committee believes that particular attention to the enhancement of satellite communications is vital, because the requirements will become more compelling as Arctic operations increase.[34]

3. *Icebreakers.* The Navy has no surface combatants hardened for ice operations. Additionally, a recent report by the National Research Council highlighted the fact that the nation has only three multimission polar icebreakers, two of which are at the end of their designed service lives.[35] One of the icebreakers is currently in caretaker status, and the operating budgets of all three are controlled by the National Science Foundation. Considering projected increases in resource development, maritime transportation, and international competition in the Arctic, U.S. icebreaking resources are clearly inadequate for meeting national needs. This deficiency is particularly significant given the recent and continuing investment in icebreaking resources by other countries, including China, Russia, Japan, South Korea, and the European Union.[36] Icebreakers are an important naval force component that will be necessary to sustaining Arctic operations. The defining of future Coast Guard icebreaker needs and investment strategy, including the feasibility of rehabilitating existing vessels, is an imperative.

The design of the next generation of surface combatants should incorporate deliberate consideration of operating in ice-covered seas. For the Navy, a recent

---

[34]Related to this, the committee reviewed information on national imaging capabilities that may become increasingly important as Arctic activities increase. Information on national imaging capabilities and the Global Fiducials Library is available at http://gfl.usgs.gov/. Accessed November 29, 2009.

[35]The three U.S. Coast Guard icebreakers are the *Polar Star,* commissioned into service in 1976; the *Polar Sea,* commissioned in 1978; and the *Healy,* commissioned in 2000. Each vessel was designed for a 30-year service life. The *Polar Star* has been in caretaker status since 2006. The *Polar Sea* is in operational condition but, because of its age, requires increasing amounts of maintenance to remain in operation. See National Research Council, 2006, *Polar Icebreakers in a Changing World: An Assessment of U.S. Needs,* The National Academies Press, Washington, D.C. See also Ronald O'Rourke, Congressional Research Service (CRS), 2009, *Coast Guard Polar Icebreaker Modernization: Background, Issues, and Options for Congress,* CRS 7-5700, RL34391, Washington, D.C., May 29.

[36]For example, a 2006 National Research Council report that lists a world inventory estimate of polar and Baltic icebreakers states that Russia has by far the largest fleet of icebreakers, although some of them are aging and some are used to keep supply lines open to Russia's Arctic coastal settlements. Data in the 2006 study indicate that Russia has 18 icebreakers, 7 of which are nuclear-powered; Finland and Sweden are reported to have 7 icebreakers each; and Canada is reported to have 6 icebreakers. See National Research Council, 2006, *Polar Icebreakers in a Changing World: An Assessment of U.S. Needs*, The National Academies Press, Washington, D.C., pp. 57-59. China, Japan, and South Korea have also made recent investments in new icebreakers targeted for polar research. For example, see "China to Build Own Icebreakers for Poles," available at http://www.shanghaidaily.com/sp/article/2009/200910/20091008/article_415716.htm. Accessed November 24, 2009.

report by the Center for Naval Analyses noted that current surface combatants might be modified or retrofitted for Arctic operations by having steel added around the waterline but that this would provide only marginal capability.[37] Ice-capable operation requires not only hull protection, but also strengthened propellers, rudders, seawater intakes, and so on. In this committee's opinion, it is better to build ice-capable ships from the ground up, either by incorporating the capability into current designs or by redesigning a new class of vessels, as Norway and Denmark are reportedly doing. It may also be wise to build more robust under-ice capability into some fraction of future Virginia-class nuclear-powered attack submarines (SSNs) to support the projected increase in under-ice missions.[38]

Importance of Strong Maritime Partnerships

The committee recognizes that policy decisions and associated trade-offs against current national defense priorities will be necessary before additional Arctic-related resources are allocated. However, if Arctic sea ice continues to retreat at a rapid pace and the Arctic region becomes truly ice-free during the summer months as predicted by the upper-end projections—such as those adopted by the Navy Task Force Climate Change[39]—the current naval asset posture may be insufficient for the U.S. Navy's maritime domain awareness strategy and insufficient to support U.S. national security interests.[40]

---

[37]Michael D. Bowes. 2009. *Impact of Climate Change on Naval Operations in the Arctic,* CAB D0020034.A3/1REV, Center for Naval Analyses, Alexandria, Va., April.

[38]Public news articles have reported that the nuclear-powered-submarine *Texas* (SSN-775) and its 134-member crew recently completed an Arctic mission. The *Texas* reportedly broke through the ice near the North Pole and stayed on the surface for 24 hours and was the third U.S. submarine to visit the region in 2009. For deployment on Arctic missions, Virginia-class attack submarines such as the *Texas* reportedly carry an "Arctic sensor suite" similar to that carried by the older Los Angeles-class submarines that have previously traversed waters near the North Pole. This sensor suite is not a built-in capability, but instead only an add-on before deploying to an Arctic region. A Navy spokesperson has been quoted as saying that "Virginia-class submarines are not ice-hardened, and there are no plans to add ice-hardening to their designs." See "Loose Cannon and Nuclear Submarines," *CanWest News Service,* November 16, 2009, and "VA-Class Submarines Carry Arctic Sensor Suite in Northern Waters," *Inside the Navy*, November 30, 2009.

[39]Throughout this report, the term "ice-free" is used to mean that sea ice is diminished to the point that ice-hardened vessels are not required for safe navigation. In the near term, ice-laden Arctic waters will continue to have an ice cover of variable thickness and duration and will continue to pose navigational hazards for non-ice-hardened vessels. The Navy Task Force Climate Change uses a projection of ice-free summer months in the Arctic by the year 2030 based on work conducted for the Department of Defense by the Oak Ridge National Laboratory using outputs from the Community Climate System Model version 3 (CCSM3). See Karsten Steinhaeuser, Esther Parish, Alex Sorokine, and Auroop R. Ganguly, 2009,"Projected State of Arctic Sea Ice and Permafrost by 2030," Oak Ridge National Laboratory, Oak Ridge, Tenn.

[40]For example, see U.S. Navy, "Maritime Domain Awareness Concept" (MDA 2007), available at http:www.navy.mil/navydata/cno/Navy_Maritime_Domain_Awareness_Concept_Final_2007.pdf. Accessed November 23, 2009. See also U.S. Navy, Marine Corps, and Coast Guard, "Cooperative

In the committee's opinion, strong maritime partnerships will be an important component of Arctic engagement. The U.S. Coast Guard reported to the committee that it is increasing its maritime partnerships with other Arctic nations,[41] and the committee sees a need for such partnerships in the Arctic region (including bilateral and multinational alliances) to be strengthened and extended for both the Navy and the Coast Guard.[42] This should involve particular attention to cooperation with Canada, with the potential for the sharing of facilities and capabilities.

There are also national and international policy implications for naval operations in the Arctic. For example, the fact that the United States has signed but not yet ratified the United Nations Convention on the Law of the Seas (UNCLOS)[43] will become even more problematic with time and as more states call for international recognition of their claims in the Arctic. The UNCLOS provides a legal framework for the settlement of such disputes.[44]

Finding 3: Activities in the Arctic sponsored by commercial enterprises and nation-states are increasing. However, neither the U.S. Navy nor the U.S.

Strategy for 21st Century Seapower," available at http:www.navy.mil/maritime/MaritimeStrategy.pdf. Accessed November 23, 2009. Also, recent news articles report that Russia's Security Council has publicly outlined plans to create a dedicated Arctic military force to protect the country's interest in the Far North. One goal of the plan is "to increase the effectiveness of cooperation with the border agencies (coast guards) of neighboring states in the fight against terrorism on the high seas, combat illegal migration and defend marine life and resources." See "Russia's New Arctic Fighting Force," available at www.wired.com/dangerroom/2009/russias-new/. Accessed November 24, 2009.

[41]CAPT James J. Fisher, USCG, Chief, Office of Policy Integration, Headquarters, "The Coast Guard Has 143 Years of Arctic Service," presentation to the committee, September 18, 2009, Washington, D.C.

[42]See National Research Council, 2008, *Maritime Security Partnerships,* The National Academies Press, Washington, D.C.

[43]The United Nations Convention on the Law of the Seas (UNCLOS) comprises 320 articles and 9 annexes governing all aspects of ocean space, including marine scientific research, commercial activities, the permissible breadth of the territorial sea (the part of the ocean nearest the shore, over which the coastal state enjoys sovereignty), and the settlement of disputes relating to ocean matters. A full profile of the UNCLOS, its origin, and its original issues is provided in a publicly available report, *Law of the Sea: The End Game, Intelligence Community Assessment,* published by the National Intelligence Council in 1996. The report is available at http://www.dni.gov/nic/special_endgame.html. Accessed November 23, 2009.

[44]U.S. Navy and U.S. Coast Guard leadership have provided public testimony on the potential value and impact of UNCLOS ratification on U.S. naval operations. For example, the congressional testimony of former Chief of Naval Operations Admiral Vernon Clark states that the Law of the Sea Convention "supports our ability to operate wherever, whenever, and however needed under the authority of widely accepted law. The Convention codifies the right to transit through, over, and under essential international [straits] and archipelagic waters. It reaffirms the sovereign immunity of our warships and other public vessels. . . . And it preserves our rights to conduct military activities and operations in exclusive economic zones without the need for permission from or prior notice to foreign governments." See Statement of Admiral Vernon Clark, U.S. Navy (Ret.), former Chief of Naval Operations, to the United States Senate Committee on Foreign Relations. Available at www.virginia.edu/colp/pdf/ClarkTestimony071004.pdf. Accessed December 14, 2009.

Coast Guard is currently well equipped for increased maritime operations in the Arctic, or for what might become contentious positioning for territorial sovereignty and for natural resources among bordering nations. This situation may pose a risk for future U.S. national security in the longer term owing to the inability of current U.S. naval assets to project a routine military presence in the region, despite the opening of new sea-lanes. Partnerships with other nations will help mitigate this risk, although the U.S. failure to ratify the United Nations Convention on the Law of the Seas (UNCLOS) exacerbates the matter.

Recommendation 3: The Chief of Naval Operations should support the initiatives of the Commandant of the U.S. Coast Guard to define future Coast Guard icebreaker needs, as well as return operational control to the Coast Guard as soon as possible. The committee sees a need in the Arctic region for increased partnerships, including bilateral and multinational alliances. These partnerships should be strengthened and extended for both the Navy and the Coast Guard. The CNO should also continue his efforts, together with other military and political leaders, to secure rapid U.S. ratification of the UNCLOS.

4. Climate-Change-Related Technical Issues Impacting Naval Operations, Particularly in the Arctic and at High Latitudes

In its initial deliberations, the committee identified four general areas of climate-change-related technical issues that may affect naval operations: antisubmarine warfare (ASW), sensors, communications, and information and charting systems. The committee's preliminary observations on these systems are offered below.

Antisubmarine Warfare

Global climate change is projected to have a growing impact on the properties and dynamics of the ocean. This committee received initial briefings on two areas associated with these changes: (1) the direct effect of changing thermal structure, ocean salinity, and acidification on the performance of acoustical sensors and torpedoes; and (2) the future viability of Navy databases that are used in tactical planning. The second of these is of more concern for naval warfighting capabilities, especially antisubmarine warfare.[45]

[45]This letter report addresses only the most immediate concerns of potential ASW implications in the Arctic. The committee has received briefings which suggest that potential increases in ocean acidification will have only minor effects on sound transmission; acidification thus does not receive expansive coverage in this letter report as an immediate item of concern. All aspects of the potential

Warming of the upper layers of the ocean produces downward-refracting acoustical conditions, which exist routinely during the summer months throughout much of the world's ocean area. These conditions normally produce shorter acoustic detection ranges, but this is nothing inherently new or climate-change-related and is not outside the operating scope of current systems. Some ocean areas—most notably the high northern latitudes of the Atlantic—may also have reduced salinity in the upper layer due to freshwater input from melting land-ice and from higher than average precipitation and runoff into rivers. This reduced salinity may also affect acoustical propagation conditions, but, similar to the warming of upper layers discussed above, it produces nothing inherently outside the operating scope of current systems. In general, the U.S. Navy needs to monitor the changes in Arctic water mass on an ongoing basis, with the monitoring supported by high-resolution bathymetric data in the Navy's databases.

According to presentations to the committee, the ocean temperature and salinity data that currently support the Navy's ASW tactical planning may need attention. Fleet ASW platforms make tactical predictions based on in situ measurements of ocean temperature versus depth, using expendable bathythermographs (XBTs). These in situ measurements of temperature are then combined with historical measurements of salinity to produce profiles of sound velocity. The temperature data can be gathered by unmanned underwater vehicles (UUVs) and thermometry systems. The profiles are then used to calculate how sound will propagate and thus to determine how best to employ acoustic sensors. An altered ocean can cause two problems with this system: (1) The in situ XBT temperature profiles are compared to climatological values contained in the Navy's historical ocean-temperature databases. The committee was informed that if this in situ measurement deviates too much from historical norms, the actual, in situ reading is disregarded or assigned a different statistical weighting.[46] (2) As mentioned above, in some areas such as the high-latitude North Atlantic, the oceans may experience reduced salinity in their upper layers. Thus, the databases with historical data on salinity used in at-sea tactical predictions may be inaccurate and could lead to incorrect ASW decisions regarding employment of the acoustical sensors.

Sensors

Global Positioning System coverage for surface navigation is slightly degraded in the high latitudes (50-ft. horizontal precision has been demonstrated

---

impact of increases in ocean acidification on naval operations, including the most recent research in this area, and broader climate change implications for ASW in the world's oceans, will be explored during the preparation of the committee's final report.

[46]Naval operators do not actually throw out XBTs based on a fixed deviation from historical climatology data. Instead, they use statistical techniques that assign error probabilities (based on climatology and other factors) to each data point in the XBT, and then weight each data point when assimilating it into an ocean model run.

at the North Pole), but this coverage is adequate for the navigational purposes of surface ships. However, due to low satellite-elevation angles, GPS altitude precision in high latitudes is substantially degraded and may adversely affect certain aircraft operations. Taken together, these conditions make precision search-and-rescue operations difficult, especially in severe weather. Also, naval airborne surveillance and surface-ship-radars operating in high latitudes may suffer degraded performance due to ionospheric activity.

As ocean-floor surveys and mineral exploration operations increase in the Arctic, accurate underwater navigation and position fixing will become increasingly important. Although relative undersea navigation techniques are well developed, absolute-reference undersea navigation is not, and it should be further developed for this region. Additionally, interest in high-accuracy bathymetric data will accelerate. Satellites now provide abundant information about the retreat of Arctic ice caps, but data on ice thickness continue to be sparse. Innovative ways to obtain that data (such as through the use of UUVs or submarine upward-looking sonar) should be explored.

Communications

In general at the present time, commercial voice and low-data-rate communications in the high latitudes are robustly supported by Iridium satellites. However, as discussed above, high-data-rate communications provided by GEO-based satellites degrade quickly above 65° latitude and therefore are not reliable. This inherent GEO limitation equally affects commercial communications and secure military communications. Reliable high-data-rate communications coverage in high latitudes should be explored further.

Information and Charting Systems

Accurate nautical charts of the Alaska region are limited. For example, the nautical charts of the region show vast areas that have never been surveyed or that have not been surveyed for years.[47] These limitations include widespread gaps in sounding and shoreline data on nautical charts of the region and 2-meter-level errors in the state geodetic positioning framework. There are also large gaps in tidal data and tidal-current-prediction coverage. These shortcomings have poten-

[47] NOAA is responsible for providing nautical charts of the Alaska region. The fundamental geospatial infrastructure that NOAA provides for the rest of the nation is lacking for Alaska and the Arctic, in particular. Alaska is the only state without digital shoreline imagery and elevation maps that meet nationally accepted standards. Also, the state's reference system has neither the density of control points to support submeter-level accuracies for surveying and positioning activities, nor vertical data coverage for the western half of the state to support the accurate determination of elevation heights. See CAPT James J. Fisher, USCG, Chief, Office of Policy Integration, Headquarters, "Waterways Management in the Arctic," communication to the committee, September 25, 2009.

tial implications for U.S. naval operations in the broader Arctic region that the committee believes should be addressed in a review of high-latitude U.S. naval research needs.

Findings and Recommendations

Finding 4: Naval operations that depend on an accurate knowledge of the ocean's properties and of atmospheric conditions could be adversely impacted, based on projections indicating that Earth is continuing to warm, if the supporting ocean and atmospheric databases are not updated. Additionally, naval sensors, communications, satellite observation, and charting systems would be significantly challenged with respect to both capacity and performance in supporting expanded levels of naval operations in high latitudes.

Recommendation 4: The Department of the Navy (the Office of the Assistant Secretary of the Navy for Research, Development and Acquisition, in conjunction with the Office of Naval Research) should begin a critical review of climate-change-related research, technology, and supporting systems—especially those related to expanded naval operations in high latitudes. In addition, the Department of the Navy and the U.S. Coast Guard should reestablish a program of routine operations in the high latitudes to develop a better understanding of the requirements for improved performance of sensors, communications, satellite communications, and information and charting systems, as well as to plan for continual awareness of the state of the Arctic through a steady focus on data gathering and supporting research and technology development.

THE WAY AHEAD

The committee plans to continue its work over the coming months to provide an expanded and more comprehensive examination of the topics covered in this letter report and to complete its final report expeditiously. Furthermore, in the preparation of its final report, the committee plans to explore additional climate-change-related topics, such as the potential impacts on the hydrological cycle and regional freshwater balances, potential changes in disease vectors and marine and terrestrial ecosystems, and the ability of the naval forces to train and operate in more extreme weather conditions—including the potential for naval vessels to experience more severe and unpredicted storms at sea. The committee will also address the general topic of risk management for naval forces and will comment on the possible benefits of applying the U.S. Navy's unique ocean and ice scientific data collection capabilities to support and enhance the understanding of potential impacts of climate change critical to national security and future naval operations.

In the committee's opinion, U.S. maritime forces are more likely than other U.S. military forces to experience more direct impacts of climate change on their operations, installations, and missions. U.S. naval leadership should thus continue to exercise a strong voice and leadership in influencing the U.S. and international military adaptive response.

ACKNOWLEDGMENTS

The committee thanks Captain Timothy Gallaudet, USN, and it also thanks Commander Esther McClure, USN, for their invaluable work as early-stage liaisons to the committee and for their input on briefings for the committee to receive. Additionally, the committee thanks members of the Chief of Naval Operations staff, N81 in particular, and the leadership of the Navy Climate Change Task Force for meeting with the committee throughout the course of the study to date.

The committee would be very happy to brief you and your staff regarding the views expressed in this letter.

Sincerely,

Frank L. Bowman

ADM Frank L. Bowman, USN (Ret.),
Co-Chair

Antonio J. Busalacchi

Dr. Antonio J. Busalacchi, Jr.,
Co-Chair

Committee on National Security Implications of Climate Change
for U.S. Naval Forces

Enclosures

A Terms of Reference
B Committee on National Security Implications of Climate Change for U.S. Naval Forces
C Acknowledgment of Reviewers
D Summary of Data-Gathering Sessions

Enclosure A
Terms of Reference

At the request of the Chief of Naval Operations, the Naval Studies Board of the National Research Council will establish a committee to study the national security implications of climate change for U.S. naval forces (i.e., the U.S. Navy, Marine Corps, and Coast Guard). Based on the Intergovernmental Panel on Climate Change assessments and other subsequent relevant literature reviewed by the committee, the study will:

1. Examine the potential impact on U.S. future naval operations and capabilities as a result of climate change (e.g., how will U.S. future naval operations be impacted and what capabilities will be needed for U.S. future naval forces as a result of climate change? This includes an assessment of the U.S. Coast Guard and Marine Corps, and where the U.S. Navy might be required to supplement or augment their capabilities).
2. Assess the robustness of the Department of Defense's infrastructure for supporting U.S. future naval operations and capabilities in the context of potential climate change impacts (e.g., are there any U.S. military installations and/or forward-deployed bases providing support to U.S. naval forces that are potentially vulnerable as a result of climate change?).
3. Determine the potential impact climate change will have on allied force operations and capabilities (e.g., are there any allies who may need U.S. naval force support as a result of climate change? Conversely, which allied force operations and capabilities may U.S. naval forces wish to leverage as a result of climate change?).
4. Examine the potential impact on U.S. future naval antisubmarine warfare operations and capabilities in the world's oceans as a result of climate change; specifically, the technical underpinnings for projecting U.S. undersea dominance in light of the changing physical properties of the oceans.

This 15-month study will produce two reports: (1) a letter report following the third full committee meeting that summarizes the immediate challenges for U.S. naval forces in addressing each of the four above areas, as well as recommends approaches to address these challenges; (2) a comprehensive report that addresses in greater depth the full terms of reference.

Enclosure B
Committee on National Security Implications of Climate Change for U.S. Naval Forces

FRANK L. BOWMAN, ADM, USN (Ret.), Strategic Decisions, LLC, *Co-Chair*
ANTONIO J. BUSALACCHI, JR., University of Maryland, *Co-Chair*
ARTHUR B. BAGGEROER, Massachusetts Institute of Technology
CECILIA M. BITZ, University of Washington
SHARON E. BURKE, Center for New American Security
RONALD FILADELFO, Center for Naval Analyses
JEFFREY M. GARRETT, RADM, USCG (Ret.), Mercer Island, Washington
HARRY W. JENKINS, JR., USMC (Ret.), Gainesville, Virginia
CATHERINE M. KELLEHER, University of Maryland and Brown University
MAHLON C. KENNICUTT II, Texas A&M University
RONALD R. LUMAN, Applied Physics Laboratory, Johns Hopkins University
W. BERRY LYONS, Ohio State University
JAMES J. McCARTHY, Harvard University
MICHAEL J. McPHADEN, National Oceanic and Atmospheric Administration
JOHN H. MOXLEY III, Solvang, California
DAVID J. NASH, RADM, USN (Ret.), Dave Nash & Associates, LLC
HEIDI C. PERRY, Charles Stark Draper Laboratory, Inc.
J. MARSHALL SHEPHERD, University of Georgia at Athens
JOHN P. STENBIT, Oakton, Virginia
CHARLES G. WALD, Gen, USAF (Ret.), Deloitte Services, LLP
DAVID A. WHELAN, The Boeing Company
CARL WUNSCH, Massachusetts Institute of Technology

Staff

CHARLES F. DRAPER, Director, Naval Studies Board
BILLY M. WILLIAMS, Study Director
RAYMOND S. WIDMAYER, Senior Program Officer
MARTA V. HERNANDEZ, Associate Program Officer
SUSAN G. CAMPBELL, Administrative Coordinator
MARY G. GORDON, Information Officer
SEKOU O. JACKSON, Senior Project Assistant
SIDNEY G. REED, JR., Consultant

Enclosure C
Acknowledgment of Reviewers

This report has been reviewed in draft form by individuals chosen for their diverse perspectives and technical expertise, in accordance with procedures approved by the National Research Council's Report Review Committee. The purpose of this independent review is to provide candid and critical comments that will assist the institution in making its published report as sound as possible and to ensure that the report meets institutional standards for objectivity, evidence, and responsiveness to the study charge. The review comments and draft manuscript remain confidential to protect the integrity of the deliberative process. We wish to thank the following individuals for their review of this report:

Robert Duce, Texas A&M University,
Florence Fetterer, University of Colorado, Boulder,
Paul G. Gaffney II, VADM, USN (Ret.), Monmouth University,
Jacques Gansler, University of Maryland,
James D. Hull, VADM, USCG (Ret.), Annapolis, Maryland,
William A. LaPlante, Johns Hopkins University, Applied Physics Laboratory,
Joseph Pedlosky, Woods Hole Oceanographic Institution, and
John E. Rhodes, LtGen, USMC (Ret.), Balboa, California.

Although the reviewers listed above have provided many constructive comments and suggestions, they were not asked to endorse the conclusions or recommendations, nor did they see the final draft of the report before its release. The review of this report was overseen by Robert A. Frosch of Harvard University. Appointed by the National Research Council, he was responsible for making certain that an independent examination of this report was carried out in accordance with institutional procedures and that all review comments were carefully considered. Responsibility for the final content of this report rests entirely with the authoring committee and the institution.

Enclosure D
Summary of Data-Gathering Sessions

The Committee on National Security Implications of Climate Change for U.S. Naval Forces first convened in September 2009 and held three full committee meetings prior to issuing this letter report. In addition to deliberating on its tasks and approach and preparing its letter report, the committee also participated in the data-gathering sessions at these meetings, which are summarized below.

- *September 17-18, 2009, in Washington, D.C.* First full committee meeting: Briefings on current climate-change and energy-related initiatives from the Navy Task Force Climate Change; the Navy Task Force Energy; the Navy Quadrennial Defense Review Integration Group; the Office of the Deputy Chief of Naval Operations for Integration of Capabilities and Resources (N81); the Office of Facilities Branch Head, U.S. Marine Corps; the Office of Environmental Management Section, Headquarters, U.S. Marine Corps; and the Office of Policy Integration, Headquarters, U.S. Coast Guard. Additionally the committee received briefings on recently completed climate-change-related reports by the Center for New American Security, the CNA Corporation, and the National Research Council.
- *October 19-20, 2009, in Washington, D.C.* Second full committee meeting: Briefings on climate-change-related national security issues, naval installation vulnerabilities, and current research activities by representatives from the National Intelligence Council, Woods Hole Oceanographic Institution, Oak Ridge National Laboratory, the Navy Task Force Climate Change, Naval Installations Command, the Office of Naval Research, the U.S. Geological Survey, the National Ice Center, the National Oceanic and Atmospheric Administration, the University of Washington, and the University of Colorado.
- *November 19-20, 2009, in Washington, D.C.* Third full committee meeting. Briefings on human dimensions, allies' perspectives, water resource issues, and maritime operational perspectives of climate change from Columbia University's Center for International Earth Science Information Network, the Pacific Institute for Studies in Development, Environment, and Security, the British Defence Staff of the United States British Embassy, the Office of the Deputy Chief of Naval Operations for Information Plans and Strategy, and the Commandant, U.S. Coast Guard.